Introduction to Petroleum Engineering

Introduction to Petroleum Engineering

James Cameron

R CALLISTO
REFERENCE

www.callistoreference.com

Callisto Reference,
118-35 Queens Blvd., Suite 400,
Forest Hills, NY 11375, USA

Visit us on the World Wide Web at:
www.callistoreference.com

ISBN: 978-1-64116-596-9 (Hardback)

Cataloging-in-Publication Data

Introduction to petroleum engineering / James Cameron.
    p. cm.
Includes bibliographical references and index.
ISBN 978-1-64116-596-9
1. Petroleum engineering. 2. Mining engineering. 3. Petroleum--Prospecting. I. Cameron, James.
TN870 .I58 2022
665.5--dc23

Table of Contents

Preface

The branch of engineering, which deals with the processes related to the production of hydrocarbons is known as petroleum engineering. These hydrocarbons could either be in the form of natural gas or crude oil. Petroleum engineering focuses on estimating the volume of hydrocarbon reservoir which can be recovered. This is done with the help of a detailed understanding of the physical behavior of water, oil and gas within porous rock at intense pressure. Some of the sub-disciplines of petroleum engineering are reservoir engineering, drilling engineering and petroleum production engineering. There are various other disciplines, which contribute knowledge to this field such as formation, evaluation, economics and artificial lift systems. Petroleum engineering is an upcoming field of science that has undergone rapid development over the past few decades. This book is a valuable compilation of topics, ranging from the basic to the most complex advancements in this field. It will serve as a valuable source of reference for graduate and postgraduate students.

Given below is the chapter wise description of the book:

Chapter 1- Petroleum refers to a yellowing-black liquid which is found in geological formations beneath the surface of the Earth. The branch of engineering which focuses on the activities related to the production of hydrocarbons, such as crude oil and natural gas, is known as petroleum engineering. This chapter will provide a brief introduction to the varied aspects of petroleum engineering.

Chapter 2- The study of the occurrence, movement, origin, exploration and accumulation of hydrocarbon fuels is termed as petroleum geology. Some of the major elements of study within this field are source rocks, reservoir rocks, trap rocks and cap rocks. This chapter has been carefully written to provide an easy understanding of these elements of petroleum geology.

Chapter 3- The search for petroleum and natural gas within the Earth by using methods and techniques of petroleum geology is called petroleum exploration. The diverse applications of petroleum exploration as well as the techniques used within it, such as well logging, have been thoroughly discussed in this chapter.

Chapter 4- The process through which usable petroleum is drawn out from beneath the surface of the Earth is called petroleum extraction. This process is divided into a few phases, namely, field appraisal phase, field development phase, petroleum production phase and well abandonment phase. This chapter discusses in detail these phases of petroleum extraction.

Chapter 5- The industrial process plant where the transformation and refinement of crude oil into gasoline, naphtha, diesel fuel, kerosene, jet fuel, liquefied petroleum gas, etc. takes place is known as a petroleum refinery. All the diverse processes which take place in a petroleum refinery have been carefully analyzed in this chapter.

Chapter 6- There are numerous products which are extracted from petroleum, such as diesel fuel, gasoline/petrol, paraffin wax, kerosene, petroleum coke, lubricating oil, liquefied petroleum gas, petroleum jelly, naphtha, etc. The topics elaborated in this chapter will help in gaining a better perspective about these petroleum products.

At the end, I would like to thank all those who dedicated their time and efforts for the successful completion of this book. I also wish to convey my gratitude towards my friends and family who supported me at every step.

James Cameron

Chapter 1

Understanding Petroleum Engineering

Petroleum refers to a yellowing-black liquid which is found in geological formations beneath the surface of the Earth. The branch of engineering which focuses on the activities related to the production of hydrocarbons, such as crude oil and natural gas, is known as petroleum engineering. This chapter will provide a brief introduction to the varied aspects of petroleum engineering.

Petroleum

Petroleum or crude oil is a naturally occurring liquid found in formations in the Earth consisting of a complex mixture of hydrocarbons (mostly alkanes) of various lengths. The approximate length range is C_5H_{12} to $C_{18}H_{38}$. Any shorter hydrocarbons are considered natural gas or natural gas liquids, while long-chain hydrocarbons are more viscous, and the longest chains are paraffin wax. In its naturally occurring form, it may contain other nonmetallic elements such as sulfur, oxygen, and nitrogen. It is usually black or dark brown (although it may be yellowish or even greenish) but varies greatly in appearance, depending on its composition. Crude oil may also be found in semi-solid form mixed with sand, as in the Athabasca oil sands in Canada, where it may be referred to as crude bitumen.

Petroleum is used mostly, by volume, for producing fuel oil and gasoline (petrol), both important "primary energy" sources. In a typical barrel 84 percent (37 of 42 gallons) of the hydrocarbons present in petroleum is converted into energy-rich fuels (petroleum-based fuels), including gasoline, diesel, jet, heating, and other fuel oils, and liquefied petroleum gas.

Ignacy Łukasiewicz - inventor of the refining of kerosene from crude oil.

Due to its high energy density, easy transportability and relative abundance, it has become the world's most important source of energy since the mid-1950s. Petroleum is also the raw material for many chemical products, including solvents, fertilizers, pesticides, and plastics; the 16 percent not used for energy production is converted into these other materials.

Petroleum is found in porous rock formations in the upper strata of some areas of the Earth's crust. There is also petroleum in oil sands. Known reserves of petroleum are typically estimated at around 1.2 trillion barrels without oil sands, or 3.74 trillion barrels with oil sands However, oil production from oil sands is currently severely limited. Consumption is currently around 84 million barrels per day, or 4.9 trillion liters per year. Because of reservoir engineering difficulties, recoverable oil reserves are significantly less than total oil-in-place. At current consumption levels, and assuming that oil will be consumed only from reservoirs, known reserves would be gone in about 32 years, around 2039, potentially leading to a global energy crisis. However, this ignores any new discoveries, changes in consumption, using oil sands, using synthetic petroleum, and other factors.

Formation

Chemistry

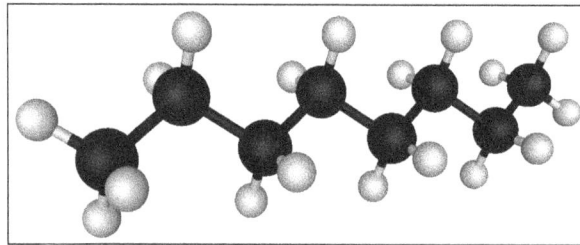

Octane, a hydrocarbon found in petroleum, lines are single
bonds, black spheres are carbon, white spheres are hydrogen.

The chemical structure of petroleum is composed of hydrocarbon chains of different lengths. These different hydrocarbon chemicals are separated by distillation at an oil refinery to produce gasoline, jet fuel, kerosene, and other hydrocarbons. The general formula for these alkanes is C_nH_{2n+2}. For example 2,2,4-trimethylpentane (isooctane), widely used in gasoline, has a chemical formula of C_8H_{18} and it reacts with oxygen exothermically:

$$2C_sH_{18(l)} + 25O_{2(g)} \rightarrow 16CO_{2(g)} + 18H_2O_{(l)} + heat$$

Incomplete combustion of petroleum or gasoline results in emission of poisonous gases such as carbon monoxide and/or nitric oxide. For example:

$$C_sH_{18(l)} + 12.5O_{2(g)} + N_{2(g)} \rightarrow 6CO2_{(g)} + 2NO_{(g)} + 9H_2O_{(l)} + heat$$

Formation of petroleum occurs in a variety of mostly endothermic reactions in high temperature and/or pressure. For example, a kerogen may break down into hydrocarbons of different lengths.

Biogenic Theory

Most geologists view crude oil and natural gas as the product of compression and heating of ancient organic materials over geological time. According to this theory, oil is formed from the preserved remains of prehistoric zooplankton and algae which have been settled to the sea (or lake) bottom in large quantities under anoxic conditions. Terrestrial plants, on the other hand, tend to form coal. Over geological time this organic matter, mixed with mud, is buried under heavy layers of sediment. The resulting high levels of heat and pressure cause the organic matter to chemically change during diagenesis, first into a waxy material known as kerogen which is found in various oil shales around the world, and then with more heat into liquid and gaseous hydrocarbons in a process known as catagenesis. Because most hydrocarbons are lighter than rock or water, these sometimes migrate upward through adjacent rock layers until they become trapped beneath impermeable rocks, within porous rocks called reservoirs. Concentration of hydrocarbons in a trap forms an oil field, from which the liquid can be extracted by drilling and pumping. Geologists often refer to an "oil window" which is the temperature range that oil forms in—below the minimum temperature oil remains trapped in the form of kerogen, and above the maximum temperature the oil is converted to natural gas through the process of thermal cracking. Though this happens at different depths in different locations around the world, a 'typical' depth for the oil window might be 4–6 km. Note that even if oil is formed at extreme depths, it may be trapped at much shallower depths, even if it is not formed there (the Athabasca Oil Sands is one example). Three conditions must be present for oil reservoirs to form: first, a source rock rich in organic material buried deep enough for subterranean heat to cook it into oil; second, a porous and permeable reservoir rock for it to accumulate in; and last a cap rock (seal) that prevents it from escaping to the surface.

The vast majority of oil that has been produced by the earth has long ago escaped to the surface and been biodegraded by oil-eating bacteria. Oil companies are looking for the small fraction that has been trapped by this rare combination of circumstances. Oil sands are reservoirs of partially biodegraded oil still in the process of escaping, but contain so much migrating oil that, although most of it has escaped, vast amounts are still present - more than can be found in conventional oil reservoirs. On the other hand, oil shales are source rocks that have never been buried deep enough to convert their trapped kerogen into oil.

The reactions that produce oil and natural gas are often modeled as first order breakdown reactions, where kerogen is broken down to oil and natural gas by a set of parallel reactions, and oil eventually breaks down to natural gas by another set of reactions. The first set was originally patented in 1694 under British Crown Patent No. 330 covering. The latter set is regularly used in petrochemical plants and oil refineries.

Abiogenic Theory

The idea of abiogenic petroleum origin was championed in the Western world by astronomer Thomas Gold based on thoughts from Russia, mainly on studies of Nikolai Kudryavtsev. The idea proposes that hydrocarbons of purely geological origin exist in the planet. Hydrocarbons are less dense than aqueous pore fluids, and are proposed to migrate upward through deep fracture networks. Thermophilic, rock-dwelling microbial life-forms are proposed to be in part responsible for the biomarkers found in petroleum.

This theory is a minority opinion, especially amongst geologists; no oil companies are currently known to explore for oil based on this theory.

Classification

The oil industry classifies "crude" by the location of its origin (e.g., "West Texas Intermediate, WTI" or "Brent") and often by its relative weight or viscosity ("light," "intermediate" or "heavy"); refiners may also refer to it as "sweet," which means it contains relatively little sulfur, or as "sour," which means it contains substantial amounts of sulfur and requires more refining in order to meet current product specifications. Each crude oil has unique molecular characteristics which are understood by the use of crude oil assay analysis in petroleum laboratories.

Barrels from an area in which the crude oil's molecular characteristics have been determined and the oil has been classified are used as pricing references throughout the world. These references are known as Crude oil benchmarks:

- Brent Crude, comprising 15 oils from fields in the Brent and Ninian systems in the East Shetland Basin of the North Sea. The oil is landed at Sullom Voe terminal in the Shetlands. Oil production from Europe, Africa and Middle Eastern oil flowing west tends to be priced off the price of this oil, which forms a benchmark.

- West Texas Intermediate (WTI) for North American oil.

- Dubai, used as benchmark for Middle East oil flowing to the Asia-Pacific region.

- Tapis (from Malaysia, used as a reference for light Far East oil).

- Minas (from Indonesia, used as a reference for heavy Far East oil).

- The OPEC Reference Basket, a weighted average of oil blends from various OPEC (The Organization of the Petroleum Exporting Countries) countries.

Means of Production

Extraction

The most common method of obtaining petroleum is extracting it from oil wells found in oil fields. After the well has been located, various methods are used to recover the petroleum. Primary recovery methods are used to extract oil that is brought to the surface by underground pressure, and can generally recover about 20 percent of the oil present. After the oil pressure has depleted to the point that the oil is no longer brought to the surface, secondary recovery methods draw another 5 to 10 percent of the oil in the well to the surface. Finally, when secondary oil recovery methods are no longer viable, tertiary recovery methods reduce the viscosity of the oil in order to bring more to the surface.

Alternative Methods

During the last oil price peak, other alternatives to producing oil gained importance. The best known such methods involve extracting oil from sources such as oil shale or tar sands. These resources are known to exist in large quantities; however, extracting the oil at low cost without negatively impacting the environment remains a challenge.

It is also possible to transform natural gas or coal into oil (or, more precisely, the various hydrocarbons found in oil). The best-known such method is the Fischer-Tropsch process. It was a concept pioneered in Nazi Germany when imports of petroleum were restricted due to war and Germany found a method to extract oil from coal. It was known as *Ersatz* ("substitute" in German), and accounted for nearly half the total oil used in WWII by Germany. However, the process was used only as a last resort as naturally occurring oil was much cheaper. As crude oil prices increase, the cost of coal to oil conversion becomes comparatively cheaper. The method involves converting high ash coal into synthetic oil in a multi-stage process. Ideally, a ton of coal produces nearly 200 liters (1.25 bbl, 52 US gallons) of crude, with by-products ranging from tar to rare chemicals.

Currently, two companies have commercialized their Fischer-Tropsch technology. Shell in Bintulu, Malaysia, uses natural gas as a feedstock, and produces primarily low-sulfur diesel fuels. Sasol in South Africa uses coal as a feedstock, and produces a variety of synthetic petroleum products.

The process is today used in South Africa to produce most of the country's diesel fuel from coal by the company Sasol. The process was used in South Africa to meet its energy needs during its isolation under Apartheid. This process has received renewed attention in the quest to produce low sulfur diesel fuel in order to minimize the environmental impact from the use of diesel engines.

An alternative method of converting coal into petroleum is the Karrick process, which was pioneered in the 1930s in the United States. It uses high temperatures in the absence of ambient air, to distill the short-chain hydrocarbons of petroleum out of coal.

More recently explored is thermal depolymerization (TDP), a process for the reduction of complex organic materials into light crude oil. Using pressure and heat, long chain polymers of hydrogen, oxygen, and carbon decompose into short-chain petroleum hydrocarbons. This mimics the natural geological processes thought to be involved in the production of fossil fuels. In theory, TDP can convert any organic waste into petroleum.

Uses

The chemical structure of petroleum is composed of hydrocarbon chains of different lengths. Because of this, petroleum may be taken to oil refineries and the hydrocarbon chemicals separated by distillation and treated by other chemical processes, to be used for a variety of purposes.

Fuels

- Ethane and other short-chain alkanes which are used as fuel,
- Diesel fuel,
- Fuel oils,
- Gasoline,
- Jet fuel,
- Kerosene,
- Liquid petroleum gas (LPG).

Other Derivatives

Certain types of resultant hydrocarbons may be mixed with other non-hydrocarbons, to create other end products:

- Alkenes (olefins) which can be manufactured into plastics or other compounds.

- Lubricants (produces light machine oils, motor oils, and greases, adding viscosity stabilizers as required).

- Wax, used in the packaging of frozen foods, among others.

- Sulfur or Sulfuric acid. These are a useful industrial materials. Sulfuric acid is usually prepared as the acid precursor oleum, a byproduct of sulfur removal from fuels.

- Bulk tar.

- Asphalt.

- Petroleum coke, used in speciality carbon products or as solid fuel.

- Paraffin wax.

- Aromatic petrochemicals to be used as precursors in other chemical production.

Consumption Statistics

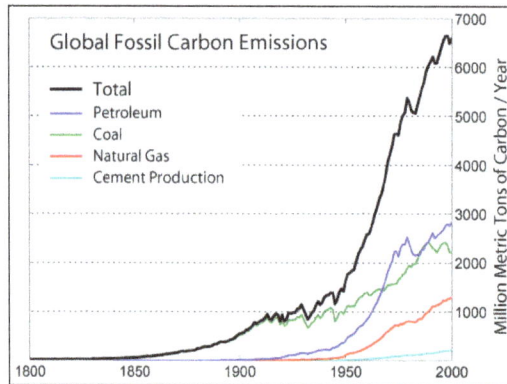

Global fossil carbon emissions, an indicator of consumption,
for 1800-2000 total is black. Oil is in blue.

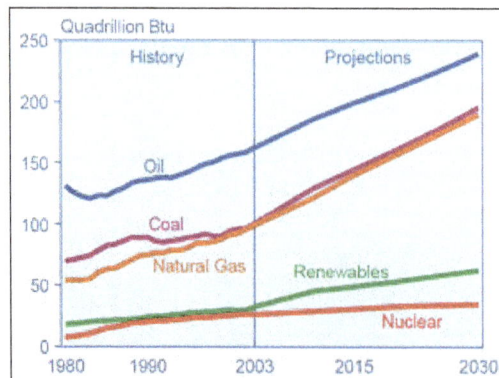

World energy consumption.

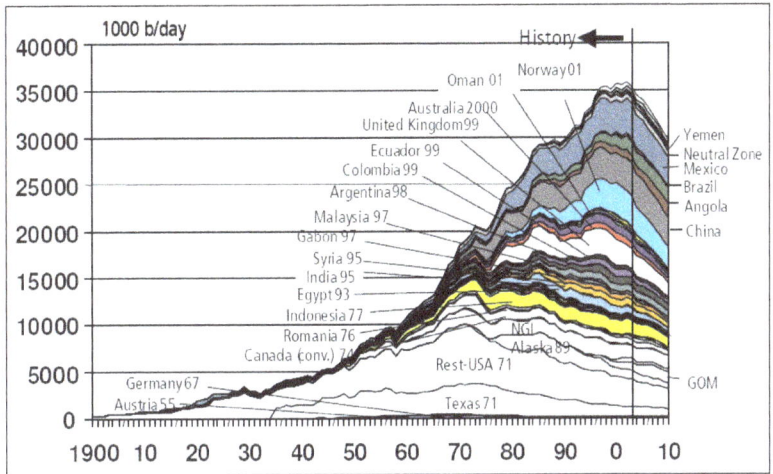

2004 U.S. government predictions for oil production other
than in OPEC and the former Soviet Union.

Environmental Effects

Diesel fuel spill on a road.

The presence of oil has significant social and environmental impacts, from accidents and routine activities such as seismic exploration, drilling, and generation of polluting wastes not produced by other alternative energies.

Extraction

Oil extraction is costly and sometimes environmentally damaging, although Dr. John Hunt of the Woods Hole Oceanographic Institution pointed out in a 1981 paper that over 70 percent of the reserves in the world are associated with visible macroseepages, and many oil fields are found due to natural leaks. Offshore exploration and extraction of oil disturbs the surrounding marine environment. But at the same time, offshore oil platforms also form micro-habitats for marine creatures. Extraction may involve dredging, which stirs up the seabed, killing the sea plants that marine creatures need to survive.

Oil spills

Volunteers cleaning up the aftermath of the Prestige oil spill.

Crude oil and refined fuel spills from tanker ship accidents have damaged natural ecosystems in Alaska, the Galapagos Islands and many other places and times in Spain (i.e. Ibiza).

Global Warming

Burning oil releases carbon dioxide into the atmosphere, which contributes to global warming. Per energy unit, oil produces less CO_2 than coal, but more than natural gas. However, oil's unique role as a transportation fuel makes reducing its CO_2 emissions a particularly thorny problem; amelioration strategies such as carbon sequestering are generally geared for large power plants, not individual vehicles.

Alternatives to Petroleum

Alternatives to Petroleum-based Vehicle fuels

The term alternative propulsion or "alternative methods of propulsion" includes both:

- Alternative fuels used in standard or modified internal combustion engines (i.e. combustion hydrogen or biofuels).

- Propulsion systems not based on internal combustion, such as those based on electricity (for example, all-electric or hybrid vehicles), compressed air, or fuel cells (i.e. hydrogen fuel cells).

Nowadays, cars can be classified between the next main groups:

- Petro-cars, this is, only use petroleum and biofuels (biodiesel and biobutanol).

- Hybrid vehicle and plug-in hybrids, that use petroleum and other source, generally, electricity.

- Petrofree car, that cannot use petroleum, like electric cars, hydrogen vehicle

Pros and Cons of Petroleum

List of the Pros of Petroleum

- Petroleum offers a stable energy resource for a variety of needs. Although we have only learned how to refine petroleum in recent generations, ancient humans were aware of this energy resource. They would caulk their ships with asphalt as far back as the 40th century BC. The hydrocarbons were used for medical purposes. The Chinese culture learned that the substance could provide light and heat, using bamboo plumbing as a way to bring it into their homes.

Our use of petroleum may have evolved over the years, but its usefulness has never changed. The amount of energy we receive through the combustion of this fossil fuel is stable and consistent. We can predict achievable outcomes based on the quality that comes from beneath the surface.

- Petroleum is what we use to create renewable energy resources. Solar energy is one of the most exciting frontiers of science because the sun provides enough resources every day to power the current needs of our planet for an entire year. If we want to create the photovoltaic panels that are necessary to harvest this resource, then we must have petroleum available for the manufacturing process. It requires more than 100L of petroleum to create a single panel, which means each one requires two years of service to repay the carbon investment.

Renewable energy resources might produce fewer emissions than the regular consumption of fossil fuels, but we would not have that option in the first place if we did not have petroleum to use.

- Petroleum provides us with a fuel which contains a high-density rating. When we create energy from a specific resource, then there is a measurement ratio which evaluates how much we receive. Petroleum has one of the highest ratings possible for a fossil fuel at 1:10,000 without releasing particles into the atmosphere. Although nuclear power is much higher with its rating, there is not the same threat of fallout or radiation when consuming this natural resource.

Because the hydrocarbons are of such an intense density level, we can turn petroleum into a variety of different fuels, consumables, and other various products. That is why it is such a valuable commodity that we consume by the millions of barrels every day.

- Petroleum is easily extracted and affordable to obtain. When we start hunting for new petroleum resources, then experts begin to look at untapped rock strata to see what might be lying beneath the surface. We then drill into this layer of rock, which creates enough pressure to force the liquid upward. Hydraulic fracturing can encourage a complete release of the reserves as well, allowing us to obtain deep wells that give us the energy we need for everyday needs.

The technologies we use today to obtain petroleum allow us to drill through the surface of the ocean too to extract what is needed. Unless the price per barrel of this commodity drops below $30, it is an affordable energy resource that we can access at almost any time with today's technologies.

- Petroleum is the foundation of our society. The infrastructures that we have across all industries and segments of society are based on how we use petroleum. By using this fuel for our daily requirements, we avoid the environmental costs of implementing new systems. There will always be some emissions that are concerning from the combustion of these fuels, so the time it would take to make up the difference between an investment in renewables and what our habits are today could be immense. That is why a slow, steady approach with a gradual transition to more renewable energy resources is the favored path to take.

- Petroleum offers numerous usage opportunities. Gasoline is arguably the most common way that we all use petroleum at some point during the day. We use this fuel to provide energy for our transportation requirements, heat for our homes, and even for the lights that we use to see at night. This product allows us to make asphalt and similar products that is useful for roadways, driveways, and sealants. Over 6,000 different products are currently manufactured thanks to the petroleum industry and how it can manipulate hydrocarbons, and more is being added to that list every day.

- Petroleum does not suffer energy loss during transportation. There are some energy resources that we use today that can lose up to half of its potential during transportation. When we are taking petroleum to a refinery to become useful, it does not experience the same outcome. We can transport this commodity over hundreds of miles through pipeline systems without changing the usefulness of the product. Tanker ships can haul millions of gallons across the vast oceans of our planet in the same way.

This advantage makes it possible to access petroleum resources in remote locations for a minimal investment, and then transport it to where our population centers are for consumption.

- Petroleum is useful as a medical product. The Ancient Chinese may have been one of the first to look at petroleum as a useful medical option, but we are continuing to walk in their footsteps. Petroleum jelly has more than 20 different medical benefits to consider by itself. Some of the anesthetics that we use today come from this fossil fuel. Dentures and denture adhesive are sometimes made with petroleum products. We even produce antihistamines and cortisone from the hydrocarbons that producers harvest from beneath the surface of our planet.

There is an excellent chance that there is more than a dozen different items that were made with petroleum in your home right now. Even some of your clothing comes from this commodity.

- Petroleum is available in different grades. Every petroleum reservoir offers unique qualities that are usable by the industry in some way. When the product contains high levels of sulfur, then the carbon-rich coke is useful in the production of synthetic crude products. The Orinoco oil sands in Venezuela and the Canada oil sands region are both examples of this version of the commodity. Although this pet-coke emits up to 80% more carbon dioxide per unit of weight, the sulfur can be refined from it so that the product can meet emissions and consumption standards that are in place around the world.

- Petroleum oil fields can go through a restoration process. When an oil well is no longer an economic resource, then it is plugged so that it can no longer be used as a commodity. When this action occurs, then the area around the well can go through a restoration process. The land can be replenished, new plants and grasses made available, and even recreational trails installed. Many facilities can go through a complete renovation in 12 months or less to become useful in other ways.

This benefit applies to offshore oil rigs as well. Some of the older platforms have been toppled into the water, creating an artificial reef that can become a new habitat for marine life. In the reef-to-rigs program, it takes about one year for nature to begin reclaiming the item with clams, sponges, coral, and other creatures covering the structure. As more life comes to the recycled rig, it can increase fish populations, create new recreational fishing opportunities, and even become a spot for diving tourism.

- Petroleum extract is a safe process. Although there have been several devastating oil spills over the past 30 years, it is essential to note that most of these are accidents that occur during the transportation of the commodity or during its extraction. The actual process is safe to the workers and the environment when it works as intended. Even when there is an incident that occurs, such as the Deep Horizon drilling rig explosion in 2010, the industry does a complete review of procedures, regulations, and technologies to ensure the problem doesn't happen again.

List of the Cons of Petroleum

- Petroleum infrastructures require continuous maintenance to continue being useful. There were over 138,000 tons of petroleum products spilled by the Sanchi oil tanker when it collided with another ship. Over 820,000 tons of petroleum were spilled into the Persian Gulf in the 1990s. Pipelines leak thousands of gallons of crude oil and other products headed for refinement each year. With the Exxon Valdez oil spill that occurred in Prince William Sound in 1989, over 37,000 metric tons of oil were spilled into the water, which was the equivalent of 260,000 barrels. More than 1,300 miles of shoreline were impacted by this event.

  Although petroleum does not lose its energy potential when we transport it, our systems are imperfect. When a spill does occur, the amount that spills forth is more than what a local habitat can generally endure. That means the quality of the soil, animal habitats, and even our livelihoods can all be adversely impacted when something goes wrong.

- Petroleum is responsible for many of our greenhouse gas emissions. We know that carbon dioxide in our atmosphere works to reflect the heat back toward our planet after the sunlight bounces off of a surface. We also know that the consumption of petroleum products either directly or through the manufacturing cycle can increase the amount of this gas that we release.

  We also know that when air is heated suddenly, it goes through an oxidation effect because of the nitrogen that is present. This circumstance creates nitrous oxide, which is also found in petroleum products in their natural state. Then this combination of factors works to change the pH balance of our atmosphere, creating the foundation for acidic precipitation. When that rainfall hits our habitats, homes, and communities, then it can negatively impact our way of life.

- Petroleum has a value which is often exploited for political gain. One could argue that many of the wars which were fought after 1950 around the world were because of access to petroleum. Even though there is an extensive export market for this commodity, the fortunes of governments rise and fall based on the available of this natural resource. If your country doesn't have it, then your government may decide that your neighbor who is rich in it looks like an inviting target.

Because the wealth that petroleum brings can change lives dramatically, the loss of its value can destroy entire nations in a heartbeat. Many of the nations in the Middle East are working to diversify their economies by 2035 to reduce their reliance on this product. Many of the financial issues facing Venezuela in 2019 are due to unanticipated shifts in commodity pricing. Because there is such a reliance on this product, it is not unusual for politicians to take advantage of this fact to destabilize regions.

- Petroleum exposure can be toxic. Did you know that petroleum exposure can be a life-threatening incident to all life on our planet? The lethality levels of this product can occur in concentrations as low as 0.4% for some forms of marine life. Even humans experience this issue, as the benzene found in this commodity is a known carcinogen. People who work to harvest this product can experience a reduction in their white blood cell counts as well, which means they can become sicker faster and easier compared to the general population.

- Petroleum changes the composition of our ocean waters. When petroleum products come into contact with marine life, it is almost guaranteed that death will occur. Even in the best-case scenario, the commodity will impede the development or behavior of the life in question. This process can change the pH levels of the ocean, which further disrupts the chain of life in the water because of increased acidity levels. Over the past two centuries, the natural pH levels of our oceans have dropped by 25% to make the water more acidic.

  Then the plastic products that we create from petroleum are a pollution hazard as well. Although we can harvest large items to turn them into parlay materials after the recycling process, microplastics are much more challenging to remove from the water. These small bits of plastic are swallowed by marine life, which then we harvest through the commercial fishing process and eat ourselves.

- Petroleum is a substance that may not last forever. The chances that we will run out of petroleum before the end of our children's lifetime is possible. There have been dire warnings in the past about this disadvantage. As far back as 1914, the U.S. Bureau of Mines stated that the world would run out of oil in just 10 years. By 1939, that number had increased to 30 years. Now we're looking at a 50-year window for this resource, with the potential to run out by 2050 to 2075.

  Is it possible that we could find new resources to continue using petroleum as we do? Absolutely – we have spent a century defeating the deadlines that experts gave us with this commodity. The definition of a fossil fuel is that it is a finite product, so even if it doesn't happen in the next few decades, there is an excellent chance that it could happen one day.

- Petroleum consumption can trigger breathing problems in some people. The consumption of petroleum products creates a higher chance of a breathing problem occurring for the individual exposed to the vapors or fumes. Even if you use a refined product for your energy requirements, the combustion process can create health issues with regular exposure. Some people can also develop an allergic reaction to this substance because we use hydrocarbons in many different products.

  Potential allergy symptoms include a runny nose, watery eyes, itching in the throat and mouth, and skin irritation. If you encounter sneezing or wheezing, then this could indicate a more severe reaction is occurring. Swelling upon contact is possible as well, which can include the air passageways.

- Petroleum can create excessive amounts of waste oil. When there is waste oil present after the petroleum is used in some way, then it contains impurities which make it challenging to recycle the product. Transmission oil, brake fluid, synthetics, and other gear box or crankcase oils all create waste products as part of their regular service. Vehicles can even drip oil out of their engines while driving, creating problems with benzene entering the water chain. There are also volatile organic compounds, or VOCs, that can create waste problems when they enter indoor spaces.

Petroleum Engineering

Petroleum engineering is the branch of engineering that focuses on processes that allow the development and exploitation of crude oil and natural gas fields as well as the technical analysis, computer modeling, and forecasting of their future production performance. Petroleum engineering evolved from mining engineering and geology, and it remains closely linked to geoscience, which helps engineers understand the geological structures and conditions favorable for petroleum deposits. The petroleum engineer, whose aim is to extract gaseous and liquid hydrocarbon products from the earth, is concerned with drilling, producing, processing, and transporting these products and handling all the related economic and regulatory considerations.

Branches of Petroleum Engineering

During the evolution of petroleum engineering, a number of areas of specialization developed: drilling engineering, production engineering and surface facilities engineering, reservoir engineering, and petrophysical engineering. Within these four areas are subsets of specialization engineers, including some from other disciplines—such as mechanical, civil, electrical, geological, geophysical, and chemical engineering. The unique role of the petroleum engineer is to integrate all the specializations into an efficient system of oil and gas drilling, production, and processing.

Drilling engineering was among the first applications of technology to oil field practices. The drilling engineer is responsible for the design of the earth-penetration techniques, the selection of casing and safety equipment, and, often, the direction of the operations. These functions involve understanding the nature of the rocks to be penetrated, the stresses in these rocks, and the techniques available to drill into and control the underground reservoirs. Because drilling involves organizing a vast array of service companies, machinery, and materials, investing huge funds, working with local governments and communities, and acknowledging the safety and welfare of the general public, the engineer must develop the skills of supervision, management, and negotiation.

The work of production engineers and surface facilities engineers begins upon completion of the well—directing the selection of producing intervals and making arrangements for various accessories, controls, and equipment. Later the work of these engineers involves controlling and measuring the produced fluids (oil, gas, and water), designing and installing gathering and storage systems, and delivering the raw products (gas and oil) to pipeline companies and other transportation agents. These engineers are also involved in such matters as corrosion prevention, well performance, and formation treatments to stimulate production. As in all branches of petroleum engineering, production engineers and surface facilities engineers cannot view the in-hole or surface processing problems in isolation but must fit solutions into the complete reservoir, well, and surface system, and thus they must collaborate with both the drilling and reservoir engineers.

Reservoir engineers are concerned with the physics of oil and gas distribution and their flow through porous rocks—the various hydrodynamic, thermodynamic, gravitational, and other forces involved in the rock-fluid system. They are responsible for analyzing the rock-fluid system, establishing efficient well-drainage patterns, forecasting the performance of the oil or gas reservoir, and introducing methods for maximum efficient production.

To understand the reservoir rock-fluid system, the drilling, production, and reservoir engineers are helped by the petrophysical, or formation-evaluation, engineer, who provides tools and analytical techniques for determining rock and fluid characteristics. The petrophysical engineer measures the acoustic, radioactive, and electrical properties of the rock-fluid system and takes samples of the rocks and well fluids to determine porosity, permeability, and fluid content in the reservoir.

While each of these four specialty areas have individual engineering responsibilities, it is only through an integrated geoscience and petroleum engineering effort that complex reservoirs are now being developed. For example, the process of reservoir characterization, otherwise known as developing a static model of the reservoir, is a collaboration between geophysicists, statisticians, petrophysicists, geologists, and reservoir engineers to map the reservoir and establish its geological structure, stratigraphy, and deposition. The use of statistics helps turn the static model into a dynamic model by smoothing the trends and uncertainties that appear in the gaps in the static model. The dynamic model is used by the reservoir engineer and reservoir simulation engineer with support from geoscientists to establish the volume of the reservoir based on its fluid properties, reservoir pressures and temperatures, and any existing well data. The output of the dynamic model is typically a production forecast of oil, water, and gas with a breakdown of the associated development and operations costs that occur during the life of the project. Various production scenarios are constructed with the dynamic model to ensure that all possible outcomes—including enhanced recovery, subsurface stimulation, product price changes, infrastructure changes, and the site's ultimate abandonment—are considered. Iterative inputs from the various engineering and geoscience team members from initial geology assessments to final reservoir forecasts of reserves being produced from the simulator help minimize uncertainties and risks in developing oil and gas.

Chapter 2

Petroleum Geology

The study of the occurrence, movement, origin, exploration and accumulation of hydrocarbon fuels is termed as petroleum geology. Some of the major elements of study within this field are source rocks, reservoir rocks, trap rocks and cap rocks. This chapter has been carefully written to provide an easy understanding of these elements of petroleum geology.

The science of geology is one of the fundamental sciences of petroleum engineering. Petroleum Geology is the study of hydrocarbon fuels that deal with the origin, occurrence and exploitation of gas and oil fields. It is basically concerned with some key elements in sedimentary basins such as source, seal, reservoir, trap, timing, etc. A scientist who works in the field of Petroleum Geology is called petroleum geologist. It showcases all aspects of discovery and production. Moreover, with thorough study the geologists are able to locate the exact position of oil deposits and lead.

Petroleum Geology plays a vital role in understanding the main geological concepts that affect reservoirs and oil fields. It refers to the certain set geological disciplines that are applied when geologists search for hydrocarbons. The chief disciplines are source rock analysis, exploration stage, basin analysis, production stage, reservoir analysis, and appraisal stage. The analysis helps in determining the permeability and porosity of the drilling samples and also helps to examine the contiguous parts of the reservoir. Petroleum Geology also used to study the sedimentary properties of the rocks.

Elements of Petroleum Geology

Source rocks, reservoir rock, seal and trap are the key elements of petroleum systems which are provided by the interpretation of data from reflection seismology and electromagnetic geophysical techniques performed in a particular geographic area. Each of these elements is evaluated in a particular way to determine the potentiality of the system.

Source Rock

Source rocks are rocks that contain sufficient organic material to create hydrocarbons when subjected to heat and pressure over time. Source rocks are usually shales or limestones (sedimentary rocks). To be a productive source rock, the rock needs time to mature (time to form the oil and/or gas) and the hydrocarbons need to be able to migrate to a reservoir or seep. Source rocks are usually a separate layer from the reservoir rock layers but occasionally they can be both source and reservoir. Source rocks are often offset from the reservoir, meaning that they are not directly below the reservoir but off to the side.

As the source rock becomes more deeply buried under layers of sediment, the temperature begins to increase triggering geochemical reactions that convert the organic materials into hydrocarbons.

Crude oil forms from 65 to 150 degrees Celsius. If the temperature goes over 150 degrees Celsius, natural gas can be formed. This usually occurs at greater depths than oil formation. Marine rocks tend to form oil while terrestrial, or land, rocks tend to form gas.

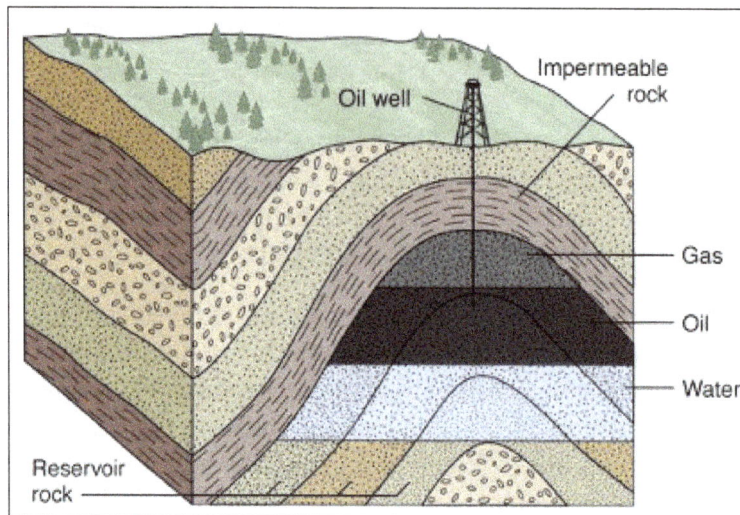

An anticline oil and gas reservoir: The source rock would
be the bottom-most layer or set off to either side.

Once the hydrocarbons have formed, they migrate from the source rock. They are then either trapped in a reservoir or are "lost" during migration. "Lost" hydrocarbons escape from the Earth through seeping (essentially they leak out of the ground) or disperse throughout the Earth's surface instead of collecting in one spot. It is estimated that only 10% of the oil and gas that has formed has been trapped. Oil and gas reservoirs typically form in traps. There are a few types of traps: anticlinal, fault, stratigraphic, and reef and/or salt traps. While the exact mechanics of each trap type differs, they all trap oil and gas in a reservoir by having an impermeable cap rock layer and by sealing the bottom of the reservoir with water (oil and gas are less dense than water) or another impermeable rock layer.

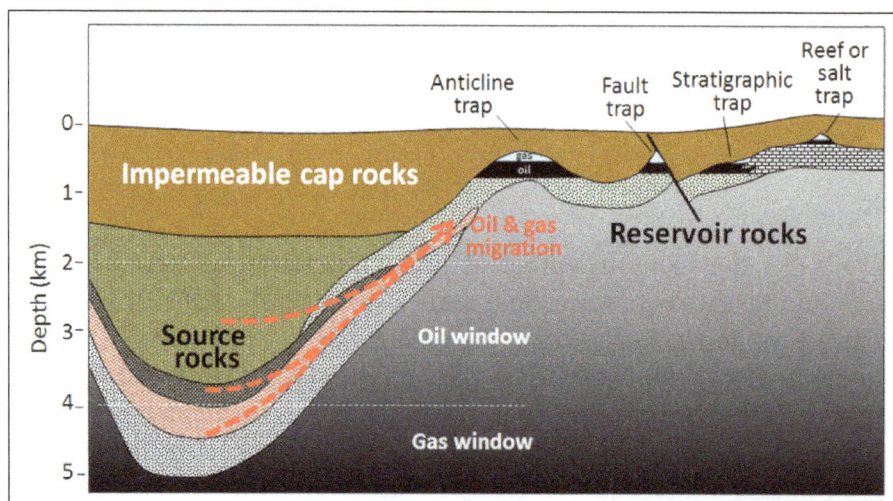

Diagram showing the windows for oil and gas formation at different depths/temperatures. Natural gas forms deeper than oil because the source rock will be at a higher temperature and pressure. After formation, these can move quite some distance before being trapped.

Types of Source Rocks

Source rocks are classified from the types of kerogen that they contain, which in turn governs the type of hydrocarbons that will be generated.

- Type I source rocks are formed from algal remains deposited under anoxic conditions in deep lakes: they tend to generate waxy crude oils when submitted to thermal stress during deep burial.

- Type II source rocks are formed from marine planktonic and bacterial remains preserved under anoxic conditions in marine environments: they produce both oil and gas when thermally cracked during deep burial.

- Type III source rocks are formed from terrestrial plant material that has been decomposed by bacteria and fungi under oxic or sub-oxic conditions: they tend to generate mostly gas with associated light oils when thermally cracked during deep burial. Most coals and coaly shales are generally Type III source rocks.

Maturation and Expulsion

With increasing burial by later sediments and increase in temperature, the kerogen within the rock begins to break down. This thermal degradation or cracking releases shorter chain hydrocarbons from the original large and complex molecules occurring in the kerogen.

The hydrocarbons generated from thermally mature source rock are first expelled, along with other pore fluids, due to the effects of internal source rock over-pressuring caused by hydrocarbon generation as well as by compaction. Once released into porous and permeable carrier beds or into faults planes, oil and gas then move upwards towards the surface in an overall buoyancy-driven process known as secondary migration.

Mapping Source Rocks in Sedimentary Basins

Areas underlain by thermally mature generative source rocks in a sedimentary basin are called generative basins or depressions or else hydrocarbon kitchens. Mapping those regional oil and gas generative "hydrocarbon kitchens" is feasible by integrating the existing source rock data into seismic depth maps that structurally follow the source horizon(s). It has been statistically observed at a world scale that zones of high success ratios in finding oil and gas generally correlate in most basin types (such as intracratonic or rift basins) with the mapped "generative depressions". Cases of long distance oil migration into shallow traps away from the "generative depressions" are usually found in foreland basins.

Besides pointing to zones of high petroleum potential within a sedimentary basin, subsurface mapping of a source rock's degree of thermal maturity is also the basic tool to identify and broadly delineate shale gas plays.

Reservoir Rock

Reservoir rocks are heterogeneous in composition and properties that are of interest to reservoir engineers. There are many types of heterogeneities encountered in geologic formations that affect the performance of the reservoir. In petroleum basins, alternating sequences of shale, sandstone, and carbonate layers are usually encountered due to the repeated encroachment or transgression of the ancient sea into the land, followed by the retreat or regression of water. The cycle may continue over a long period of time leading to the formation of many distinct geologic strata. The grading of rock grains in the vertical direction indicates the transgression/regression cycle. Transgression of the sea is associated with the finer grains deposited upward in a geologic bed or layer. Conversely, coarser grains deposited in the upward direction indicate an environment where the sea has regressed.

A geologic formation may exhibit facies change where the composition of rock may change from one rock type to another. For example, formations in many petroleum reservoirs transition from sand to predominantly shale in the lateral direction. Facies change is an indicator for a change in the depositional environment. Typically, sand particles are deposited in shallow water or coastal environments while silt and clay are deposited in lakes and relatively deep waters. Again, marine organisms can be deposited in deep seas. The occurrence of facies change in a geologic formation may define a boundary for fluid flow and affect the performance of reservoirs.

During transportation, the smaller sized sediments travel longer, and are deposited only in a very low energy environment such as deep sea. Again, well-sorted grains in rock, where most grains are of similar size, indicate long transport of sediments by water or other agents. Size and sorting of grains in reservoir rocks significantly influence the characteristics of reservoirs, including porosity and permeability of rock, which in turn affect the storage and flow capacity of petroleum.

Properties

The reservoir rock properties that are of most interest to development geologists and reservoir engineers (amongst others) are Porosity, Compressibility, and Permeability. Porosity is a rock property that defines the fraction of the rock volume that is occupied by the pore volume. Compressibility is the rock property that governs the relative change in the pore volume when pressure is either increased or decreased. Finally, permeability is the rock property that is a measure of the ease (or difficulty) with which liquids or gases can flow through a porous medium.

Rock Porosity

Porosity, ϕ, is the fraction of the Bulk Rock Volume, v_b, that is occupied by the Pore Volume, v_p. Mathematically, it is defined as:

$$\phi = \frac{V_p}{V_b}$$

The bulk volume, V_b, can also be defined as the sum of the volumes of the two constituents of the rock, pore volume and *Grain Volume*, V_g. That is:

$$V_b = V_p + V_g$$

From these two expressions, we can develop several equivalent definitions for porosity:

$$\phi = \frac{V_p}{V_b} = \frac{V_b - V_g}{V_b}$$

Figure (below) shows the porosities for three different idealized grain packings. In this figure, the rock grains are the spheres and the pore volume is the space between the spheres. Note that the porosities of these three systems are independent on the radii of the grains. In a real rock, the grains would be angular (if quartz grains in sandstones) and some of the void space might be filled with smaller grains, mineral cementation, and clays.

Porosities for Three Different Idealized Grain Packings.	
Regular Cubic-Packed Spheres	$\phi = \dfrac{V_b - V_g}{V_b} = 1 - \dfrac{\pi}{6} = 0.467$ $V_b = (2r)^3$ $V_g = 8\left[\dfrac{1}{8}\left(\dfrac{4}{3}\pi r^3\right)\right] = \dfrac{4}{3}\pi r^3$
Regular Orthorhombic-Packed Spheres	$\phi = \dfrac{V_b - V_g}{V_b} = 1 - \dfrac{\pi r^3}{3\sqrt{3}r^3} = 0.395$ $V_b = (2r)(2r)[r\,\sin(60)] = 4\sqrt{3}r^3$ $V_g = \dfrac{4}{3}\pi r^3$
Regular Rhombohedral-Packed Spheres	$\phi = \dfrac{V_b - V_g}{V_b} = 1 - \dfrac{4\pi r^3}{3\sqrt{3}r^3} = 0.260$ $V_b = (2r)(2r)(h) = 4\sqrt{2}r^3$ $h = \sqrt{4r^2 - 2r^2} = \sqrt{2}r$ $V_g = \dfrac{4}{3}\pi r^3$

At this point, we must distinguish between two types of porosity in rock: Total Porosity, φ_t, and Effective Porosity, φ_e. Real reservoir rock is comprised of connected pores and isolated pores. As the name implies, connected pores are pores which are connected to other pores and are capable of transmitting fluids. Isolated pores are pores, or groups of pores, which do not connect to other pores or form dead-end pathways. These isolated pores are incapable of transmitting fluids. Total porosity is defined as the total pore space (connected plus isolated pores) divided by the bulk volume, while effective porosity is the connected pore volume divided by the bulk volume.

Reservoir engineers are concerned with how fluids are stored and flow in the reservoir. Consequently, reservoir engineers are more concerned with effective porosity, φ_e.

The porosity is of primary concern to geologists and reservoir engineers because it is a direct measurement of the ability of a reservoir rock to store fluids. There are two ways that oil and gas professionals can obtain measurements of porosity through laboratory measurement and through the use of Well Logs.

Cap Rocks or Seal

A cap rock, or seal, provides a barrier to the migration of fluid or gas out of intended trap due to its low permeability, high capillary-entry pressure nature. For a long time, the only force causing the movement of oil and gas in the subsurface was believed to be buoyancy. If so, then to form oil and gas accumulation, their migration paths must have been stopped by a roof, i.e., caprock (seal). Clays, shales, carbonates, evaporites, and their combinations can form caprocks. The same rocks react differently to different fluids. In some cases, rocks serve as satisfactory or good conduits for water, but form barriers for oil or gas movement. In some other situations rocks yield oil but stop gas movement, etc. This is determined by capillary forces, the magnitude of which depends on fluid and rock properties (fluid density, fluid viscosity, rock structure, rock wettability) and pore size (capillary forces almost disappear when the pore diameter exceeds 0.5mm). All aforementioned rock and fluid properties are strongly affected by the subsurface temperature and pressure and geochemical environment. Caprock is a rock that prevents the flow of a given fluid at a certain temperature and pressure and geochemical conditions. Therefore, the necessary properties of a rock to act as a seal will be different for different fluids. The same rock with different fluids may or may not have sealing properties up to a complete inversion (caprock - reservoir). The caprocks can be categorized into three types.

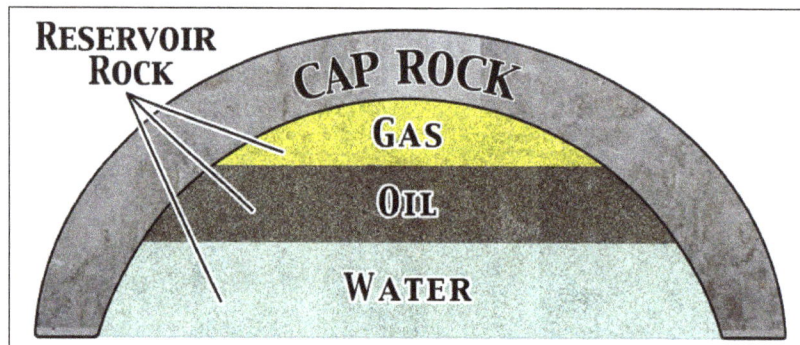

Type I caprocks are typical for argillaceous sequences in a state of continuing compaction; they are developed in areas of young subsidence of Earth's crust, with abnormally high pore water pressure. Sealing properties of these rocks are determined by the amount of capillary pressure at the contact of the reservoir and caprock, the pore pressure of water saturating the caprock, initial pressure gradient of water and the variation of hydraulic forces in the section. Oil and gas accumulations have higher potential energy than that of the formation water. These accumulations can be stable only if this energy is equal to or less than the caprock breakthrough energy. Pore water pressure in compacting argillaceous beds is always greater than the pressure in the adjacent reservoir beds. As a result, sealing capability of the Type I caprocks is determined by hydraulic sealing, by the amount of capillary pressure, and by the pressure at which water begins to flow through caprocks. Just the capillary pressure alone in such caprocks may exceed 100kg/cm2. This means that the Type I caprocks is capable of confining an oil accumulation having almost any column height. It appears that sealing capability of argillaceous caprocks does not depend on their thickness describes only the aforementioned caprock type.

Type II caprocks are associated with rocks compacted beyond the plasticity limit and having lost ability to swell on contact with water. Such rocks do not contain swelling clay minerals, and interstitial water contains surfactants. Consequently, pore water in these rocks does not have initial pressure gradient. This type of caprocks is encountered mostly in the Paleozoic and Mesozoic sediments of young and old platforms. There are no clear-cut overpressure environments there, but there is a relatively clear hydrodynamic subdivision in the section. the hydrodynamic environment may improve or lower the sealing capability of caprocks. In an extreme case, the water potential in the reservoir may exceed the water potential of the bed overlying the caprock by the value of capillary pressure. In such a situation, the caprock will be open for the vertical flow of hydrocarbons, and the trap will not exist even when potential distribution in the reservoir bed is favorable.

Type III caprocks are typical for rocks with a rigid matrix and intense fracturing. Such caprocks are mainly developed over the old platforms in regions of low tectonic mobility, with no detectable hydrodynamic breakdown of the section. Formation water potential in such regions is practically equal throughout the section and corresponds to the calculated hydrostatic potential.

The correlation between clay mineralogy and their sealing properties are as follows "The permanency in the composition of the silicate layer is a characteristic of the kaolinite group minerals. As a result, replacements within the lattice are very rare and the charges within a layer are compensated. The connection between silicate layers in the C-axis direction is implemented through hydrogen atoms, which prevents the lattice from expanding, ruling out the penetration of water and polar organic liquids. The silicate layer in the montmorillonite mineral group is variable due to a common isomorphic replacement in octahedral and narrower tetrahedral sheets. This replacement results in the disruption of the lattice neutrality. Extra charge that occurs with such replacements is compensated by exchange ions. Ion properties that maintain lattice neutrality in montmorillonite minerals (valence, size of the ion radius, polarization, etc.) define the capability of the lattice to expand along the C-axis. As a result, water and polar organic liquids can penetrate the interlayer spaces. This, in turn, leads to an increase in the volume, which drastically lowers permeability and some other properties, but at the same time improves sealing capabilities. The silicate layer of the illite mineral group is similar to the montmorillonite one. However, the excessive negative charge of the lattice is due mainly to the isomorphic replacements within tetrahedral sheets. The proximity between the source of negative charge and basal surfaces causes a stronger connection between the silicate layers of illite group compared to montmorillonite's."

Admixture of sand and silt degrades the sealing properties of clays. Especially important are the textural changes due to this admixture. Not only the mineral composition of a rock and organic matter content, but also the pore water are important in forming the major sealing properties of clays, such as degree of swelling and compressibility. The relatively low-temperature pore water is retained in argillaceous rocks up to a temperature of 100C to 150C. The temperature of water removal is higher when the concentration of dissolved components is higher. Pore water is located within pores of argillaceous rocks, and at the surfaces and along the edges of individual microblocks and microaggregates that comprise clays. The interlayer water causes swelling in montmorillonites and in degraded illites. The order in water molecules positioning, relative to the clay mineral blocks and aggregates, is rapidly altered with an increase in distance between these blocks and aggregates. Thus, a very important information for the evaluation of the role water plays in the formation of sealing properties is the knowledge of the structural status of the layer in an immediate contact with the particles surface, and the role the cations having different charge density play in the preservation of water molecules structure.

Exchange ions play a leading role in the formation of "water clouds" around microaggregates and microblocks of montmorillonite minerals and an insignificant role, with kaolinite minerals. The role played by the illite group minerals occupies an intermediate position. Carbonates cap-rocks include micro- and fine-grained, massive and laminated limestones. Almost all limestones are dolomitized to some extent and are subject to fracturing. This adversely affects their sealing properties. Carbonates with a substantial clay content have laminated texture. As a rule, this results in a deterioration rather than an improvement of sealing properties due to the emergence of weakness zones at the contact between different lithologies. Evaporite seals, which are common, include salt, anhydrite, and sometimes shales. It is a common (and probably erroneous) belief that such seals are the best and most reliable. Brittleness of these rocks at the surface conditions contradicts that belief. Besides, cores recovered in the Dnieper-Donets Basin and North Caspian Basin display macro- and microscopic fractures, which sometimes cut monolithic salt crystals. The fractures may be healed by secondary salt, but often contain traces of oil and sometimes gas bubbles. Sometimes core samples are completely saturated with oil. Permeability measured at the surface conditions can reach 100–150mD and even higher. It was established, however, that these rocks easily become plastic even at a relatively low hydrostatic or, even, uni-axial pressure (0100MPa) and the properties change with temperature. Some people considered plasticity as an important sealing property. In this connection, they believe that salt has the best sealing properties. They also believe that the reliability of caprock is not directly related to its thickness. Thus, properties of evaporites as seals change widely during the catagenesis (and in time). Similar changes also affect the other types of seals albeit not so obviously. Inclusions, such as organic matter, silt, clay or carbonate particles degrade sealing properties of evaporites due to the formation of zones of weakness around such inclusions. A careful study of numerous logs from Dnieper-Donets Basin showed the presence of clay interbeds between the top of accumulation and the evaporite sequence in all cases. It appears that these interbeds in most cases act as caprock.

Trap or Trap Rock

Trap rock is used in construction applications as concrete, macadam, and paving stones. It is also used in more technical applications as ballast for railroad track bed, ad hydraulic engineering rock for paving embankments, and more.

Two types of petroleum traps are; structural and stratigraphic. Structural traps are formed by deformation of reservoir rock, such as by folding or faulting. Stratigraphic traps are formed by deposition of reservoir rock, such as river channel or reef, or by erosion of reservoir rock, such as an angular unconformity.

Structural Trap

Structural trap is a type of geological trap that forms as a result of changes in the structure of the subsurface, due to tectonic, diapiric, gravitational and compactional processes. These changes block the upward migration of hydrocarbons and can lead to the formation of a petroleum reservoir.

Structural traps are the most important type of trap as they represent the majority of the world's discovered petroleum resources. The three basic forms of structural traps are the anticline trap, the fault trap and the salt dome trap.

Anticlinal (Fold) Trap

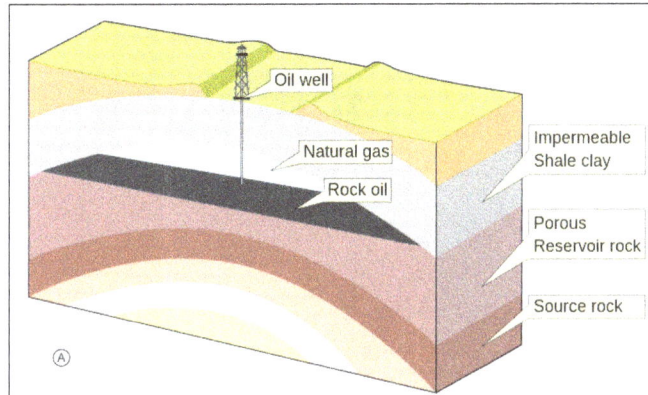

Anticlinal trap.

An anticline is an area of the subsurface where the strata have been pushed into forming a domed shape. If there is a layer of impermeable rock present in this dome shape, then hydrocarbons can accumulate at the crest until the anticline is filled to the spill point – the highest point where hydrocarbons can escape the anticline. This type of trap is by far the most significant to the hydrocarbon industry. Anticline traps are usually long oval domes of land that can often be seen by looking at a geological map or by flying over the land.

Fault Trap

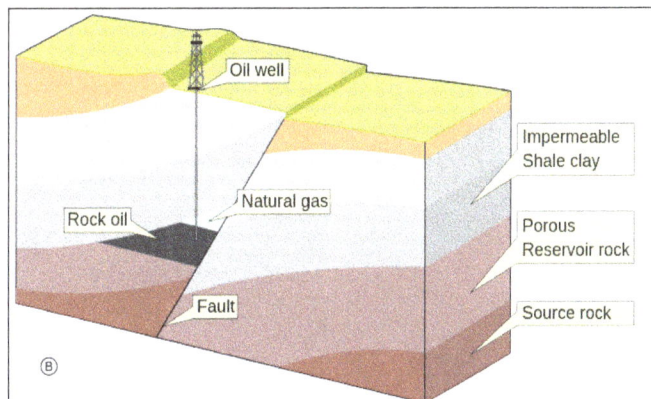

Fault trap.

This trap is formed by the movement of permeable and impermeable layers of rock along a fault line. The permeable reservoir rock faults such that it is now adjacent to an impermeable rock, preventing hydrocarbons from further migration. In some cases, there can be an impermeable substance smeared along the fault line (such as clay) that also acts to prevent migration. This is known as clay smear.

Salt Dome Trap

Masses of salt are pushed up through clastic rocks due to their greater buoyancy, eventually breaking through and rising towards the surface. This salt is impermeable and when it crosses a layer of permeable rock, in which hydrocarbons are migrating, it blocks the pathway in much the same manner as a fault trap. This is one of the reasons why there is significant focus on subsalt imaging, despite the many technical challenges that accompany it.

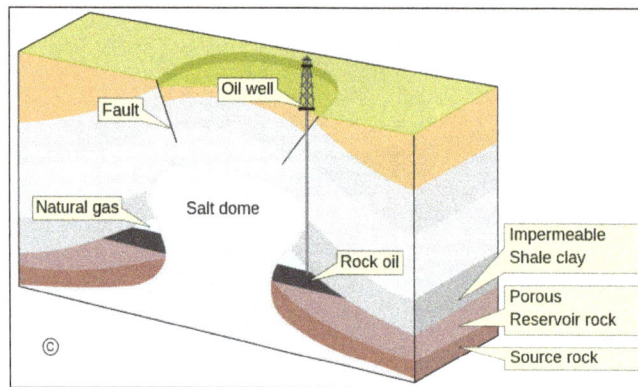

Salt Dome Trap.

Stratigraphic Traps

Stratigraphic traps are formed as a result of lateral and vertical variations in the thickness, texture, porosity or lithology of the reservoir rock. Examples of this type of trap are an unconformity trap, a lens trap and a reef trap.

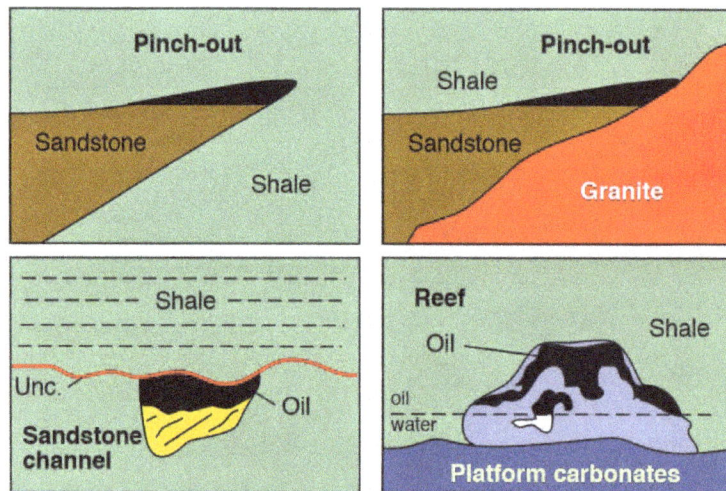

Examples of stratigraphic traps.

Two Main Groups can be Recognized

Primary

Stratigraphic traps result from variations in facies that developed during sedimentation. These include features such as lenses, pinch-outs, and appropriate facies changes.

Secondary

Stratigraphic traps result from variations that developed after sedimentation, mainly because of diagenesis. These include variations due to porosity enhancement by dissolution or loss by cementation.

Paleogeomorphic traps are controlled by buried landscape. Some are associated with prominences (hills); others with depressions (valleys). Many are also partly controlled by unconformities so are also termed unconformity traps.

Petroleum Reservoirs

Petroleum is found in underground pockets called reservoirs. Deep beneath the Earth, pressure is extremely high. Petroleum slowly seeps out toward the surface, where there is lower pressure. It continues this movement from high to low pressure until it encounters a layer of rock that is impermeable. The petroleum then collects in reservoirs, which can be several hundred meters below the surface of the Earth.

Petroleum can be contained by structural traps, which are formed when massive layers of rock are bent or faulted (broken) from the Earth's shifting landmasses. Oil can also be contained by stratigraphic traps. Different strata, or layers of rock, can have different amounts of porosity. Crude oil migrates easily through a layer of sandstone, for instance, but would be trapped beneath a layer of shale.

A petroleum reservoir or oil and gas reservoir is a subsurface pool of hydrocarbons contained in porous or fractured rock formations. Petroleum reservoirs are broadly classified as conventional and unconventional reservoirs. In case of conventional reservoirs, the naturally occurring hydrocarbons, such as crude oil or natural gas, are trapped by overlying rock formations with lower permeability. While in unconventional reservoirs the rocks have high porosity and low permeability which keeps the hydrocarbons trapped in place, therefore not requiring a cap rock. Reservoirs are found using hydrocarbon exploration methods.

Petroleum reservoirs are normally found in sedimentary rocks. Rarely, petroleum is found in fractured igneous or metamorphic rocks. Igneous and metamorphic rocks originate in high pressure and temperature conditions that do not favor the formation of petroleum reservoirs. They also do not ordinarily have the interconnected pore space needed to form a conduit for petroleum to flow to a wellbore. While a metamorphic rock may have originated as a piece of sandstone, it has been subjected to heat and pressure. Any petroleum fluid that might once have occupied the pores is cooked away. Sedimentary rocks are not the only element needed for a productive petroleum reservoir.

Several key ingredients must be present for a hydrocarbon reservoir to develop. First, a source for the hydrocarbon must be present. It is most commonly thought that oil forms from the remains of single celled aquatic life. When the remains are baked as they become buried, oil and gas are formed. As with baking a cake, the mixture can be underdone or overcooked. In either case, the formation of oil will not be optimal. Second, there needs to be a conduit from the source rock to a reservoir rock. Typically, hydrocarbons are formed in rocks which are not very amenable to modern production techniques. For hydrocarbons to be producible, they must be able to flow into wells. The flow rate must be large enough to make the wells economically viable. Two important characteristics control the economical viability of the reservoir: porosity and permeability. Porosity is the ratio of the volume of void space to the total volume. Porosity is a factor that defines the capacity of the reservoir to store fluid. Permeability is a measure of the ability of fluids to flow.

Once hydrocarbon fluid is in a suitable reservoir rock, a trapping mechanism becomes important. If the hydrocarbon fluid is not stopped from migrating, buoyancy and other forces will cause it to move toward the surface. Overriding all of these factors is timing. Without a correct timing sequence, nothing can happen. A source rock can provide billions of barrels of oil to a reservoir, but if the trap does not form until a million years after the oil has passed through the reservoir, not much will be found except perhaps an oil-stained rock.

Formation of Petroleum Reservoirs

The formation of petroleum reservoirs occurs over millions of years as oil and gas accumulations develop in underground traps formed by structural or stratigraphic geological features. These accumulations usually occur in the more porous and permeable sedimentary rock, where the petroleum molecules seep into the small inter-granular spaces, or in joints and fractures of the rock. The reservoir is the portion of the formation containing oil and gas that is hydraulically connected.

Conventional reservoirs need a source rock, a migration path, a reservoir rock and a cap rock.

The source rock is a rock that is rich in organic matter. These form as algae or other plant and animal life die and are buried in the sand. Over millions of years, these rock formations are covered by other rock formations and the pressure and temperature of the rock increases. The organic matter converts to oil and gas as it is heated.

As the oil and gas forms in the source rock, it begins to migrate through the inter-granular spaces, the joints and the fractures of the rock. The oil and gas molecules are less dense than the rock and water and naturally migrate slowly upward towards the surface.

The molecules continue migrating until they hit cap rock. Cap rock is typically a layer of impermeable rock that lacks the inter-granular spaces the molecules need to continue traveling. It overlays the reservoir rock and forms a trap. At this point, with nowhere for the molecules to go, and more molecules continuing to migrate out of the source rock, the molecules begin to accumulate in the rock directly below the cap rock – the reservoir rock.

If there is no cap rock, the molecules travel all the way to the surface, resulting in a natural oil seep.

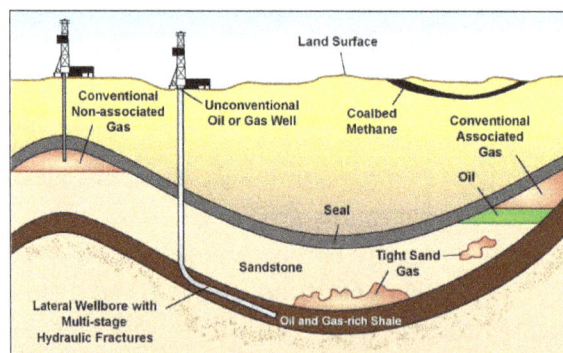

Geology for petroleum exploration, drilling and production.

In recent years, unconventional reservoirs have been developed. These unconventional reservoirs are unique in that the source rock is the reservoir rock. No migration path or cap rock is needed, as the oil and gas are produced directly from the source rock where it was generated.

Oil Field

An oil field is a land area with an abundance of oil wells extracting petroleum (crude oil) from below ground. Because oil reservoirs typically extend over a large area, possibly several hundred kilometres across, full exploitation entails multiple wells scattered across the area. In addition, there may be exploratory wells probing the edges, pipelines to transport the oil elsewhere, and support facilities.

An oil field with dozens of wells. This is the Summerland
Oil Field, near Santa Barbara.

Because an oil field may be remote from civilization, establishing a field is often an extremely complicated exercise in logistics. This goes beyond requirements for drilling, to include associated infrastructure. For instance, workers require housing to allow them to work onsite for months or years. In turn, housing and equipment require electricity and water. In cold regions, pipelines may need to be heated. Also, excess natural gas may be burned off if there is no way to make use of it—which requires a furnace, chimney and pipes to carry it from the well to the furnace.

Oil Field Mittelplate in the North Sea.

Thus, the typical oil field resembles a small, self-contained town in the midst of a landscape dotted with drilling rigs or the pump jacks, which are known as "nodding donkeys" because of their bobbing arm. Several companies, such as Hill International, Bechtel, Esso, Weatherford International, Schlumberger Limited, Baker Hughes and Halliburton, have organizations that specialize in the large-scale construction of the infrastructure and providing specialized services required to operate a field profitably.

Eagle Ford Shale flares visible from space (green and infrared wavelengths),
in the arc between "1" and "2", amid cities in southeast Texas in 2012.

More than 40,000 oil fields are scattered around the globe, on land and offshore. The largest are the Ghawar Field in Saudi Arabia and the Burgan Field in Kuwait, with more than 60 billion barrels (9.5×10^9 m³) estimated in each. Most oil fields are much smaller.

In the modern age, the location of oil fields with proven oil reserves is a key underlying factor in many geopolitical conflicts.

The term "oilfield" is also used as a shorthand to refer to the entire petroleum industry. However, it is more accurate to divide the oil industry into three sectors: upstream (crude production from wells and separation of water from oil), midstream (pipeline and tanker transport of crude) and downstream (refining, marketing of refined products and transportation to Oil stations).

Gas Field

Natural gas originates by the same geological thermal cracking process that converts kerogen to petroleum. As a consequence, oil and natural gas are often found together. In common usage, deposits rich in oil are known as oil fields, and deposits rich in natural gas are called natural gas fields.

location of Gas Fields of Iran.

In general, organic sediments buried in depths of 1,000 m to 6,000 m (at temperatures of 60 °C to 150 °C) generate oil, while sediments buried deeper and at higher temperatures generate natural gas. The deeper the source, the "drier" the gas (that is, the smaller the proportion of condensates in the gas). Because both oil and natural gas are lighter than water, they tend to rise from their sources until they either seep to the surface or are trapped by a non-permeable stratigraphic trap. They can be extracted from the trap by drilling.

Vučkovec Gas Field facility, Croatia.

The largest natural gas field is South Pars/Asalouyeh gas field, which is shared between Iran and Qatar. The second largest natural gas field is the Urengoy gas field, and the third largest is the Yamburg gas field, both in Russia.

Like oil, natural gas is often found underwater in offshore gas fields such as the North Sea, Corrib Gas Field off Ireland, and near Sable Island. The technology to extract and transport offshore natural gas is different from land-based fields. It uses a few, very large offshore drilling rigs, due to the cost and logistical difficulties in working over water.

The drillship Discoverer Enterprise is shown in the background, at work during exploratory phase of a new offshore field. The Offshore Support Vessel Toisa Perseus is shown in the foreground, illustrating part of the complex logistics of offshore oil and gas exploration and production.

Rising gas prices in the early 21st century encouraged drillers to revisit fields that previously were not considered economically viable. For example, in 2008 McMoran Exploration passed a drilling depth of over 32,000 feet (9754 m) (the deepest test well in the history of gas production) at the Blackbeard site in the Gulf of Mexico. Exxon Mobil's drill rig there had reached 30,000 feet by 2006 without finding gas, before it abandoned the site.

Estimating Reserves

After the discovery of a reservoir, a petroleum engineer will seek to build a better picture of the accumulation. In a example of a uniform reservoir, the first stage is to conduct a seismic survey to determine the possible size of the trap. Appraisal wells can be used to determine the location of oil–water contact and with it the height of the oil bearing sands. Often coupled with seismic data, it is possible to estimate the volume of an oil-bearing reservoir.

The next step is to use information from appraisal wells to estimate the porosity of the rock. The porosity, or the percentage of the total volume that contains fluids rather than solid rock, is 20–35% or less. It can give information on the actual capacity. Laboratory testing can determine the characteristics of the reservoir fluids, particularly the expansion factor of the oil, or how much the oil expands when brought from the high pressure and high temperature of the reservoir to a "stock tank" at the surface.

With such information, it is possible to estimate how many "stock tank" barrels of oil are located in the reservoir. Such oil is called the stock tank oil initially in place (STOIIP). As a result of studying factors such as the permeability of the rock (how easily fluids can flow through the rock) and possible drive mechanisms, it is possible to estimate the recovery factor, or what proportion of oil in place can be reasonably expected to be produced. The recovery factor is commonly 30–35%, giving a value for the recoverable resources.

The difficulty is that reservoirs are not uniform. They have variable porosities and permeabilities and may be compartmentalised, with fractures and faults breaking them up and complicating fluid flow. For this reason, computer modeling of economically viable reservoirs is often carried out. Geologists, geophysicists and reservoir engineers work together to build a model which allows simulation of the flow of fluids in the reservoir, leading to an improved estimate of the recoverable resources.

Reserves are only the part of those recoverable resources that will be developed through identified and approved development projects. Because the evaluation of "Reserves" has a direct impact on the company or the asset value, it usually follows a strict set of rules or guidelines (even though loopholes are commonly used by companies to inflate their own share price). The most common guidelines are the SPE PRMS guidelines, the SEC Rules, or the COGE Handbook. Government may also have their own systems, making it more complicated for investors to compare one company with another.

Production

To obtain the contents of the oil reservoir, it is usually necessary to drill into the Earth's crust, although surface oil seeps exist in some parts of the world, such as the La Brea tar pits in California, and numerous seeps in Trinidad.

Drive Mechanisms

A virgin reservoir may be under sufficient pressure to push hydrocarbons to surface. As the fluids are produced, the pressure will often decline, and production will falter. The reservoir may respond to the withdrawal of fluid in a way that tends to maintain the pressure. Artificial drive methods may be necessary.

Solution-gas Drive

This mechanism (also known as depletion drive) depends on the associated gas of the oil. The virgin reservoir may be entirely liquid, but will be expected to have gaseous hydrocarbons in solution due to the pressure. As the reservoir depletes, the pressure falls below the bubble point, and the gas comes out of solution to form a gas cap at the top. This gas cap pushes down on the liquid helping to maintain pressure.

This occurs when the natural gas is in a cap below the oil. When the well is drilled the lowered pressure above means that the oil expands. As the pressure is reduced it reaches bubble point and subsequently the gas bubbles drive the oil to the surface. The bubbles then reach critical saturation and flow together as a single gas phase. Beyond this point and below this pressure the gas phase flows out more rapidly than the oil because of its lowered viscosity. More free gas is produced and eventually the energy source is depleted. In some cases depending on the geology the gas may migrate to the top of the oil and form a secondary gas cap.

Some energy may be supplied by water, gas in water, or compressed rock. These are usually minor contributions with respect to hydrocarbon expansion.

By properly managing the production rates, greater benefits can be had from solution-gas drives. Secondary recovery involves the injection of gas or water to maintain reservoir pressure. The gas/oil ratio and the oil production rate are stable until the reservoir pressure drops below the bubble point when critical gas saturation is reached. When the gas is exhausted, the gas/oil ratio and the oil rate drops, the reservoir pressure has been reduced and the reservoir energy exhausted.

Gas Cap Drive

In reservoirs already having a gas cap (the virgin pressure is already below bubble point), the gas cap expands with the depletion of the reservoir, pushing down on the liquid sections applying extra pressure.

This is present in the reservoir if there is more gas than can be dissolved in the reservoir. The gas will often migrate to the crest of the structure. It is compressed on top of the oil reserve, as the oil is produced the cap helps to push the oil out. Over time the gas cap moves down and infiltrates the oil and eventually the well will begin to produce more and more gas until it produces only gas. It is best to manage the gas cap effectively; that is, placing the oil wells such that the gas cap will not reach them until the maximum amount of oil is produced. Also a high production rate may cause the gas to migrate downward into the production interval. In this case over time the reservoir pressure depletion is not as steep as in the case of solution based gas drive. In this case the oil rate will not decline as steeply but will depend also on the placement of the well with respect to the gas cap.

As with other drive mechanisms, water or gas injection can be used to maintain reservoir pressure. When a gas cap is coupled with water influx the recovery mechanism can be highly efficient.

Aquifer Water Drive

Water (usually salty) may be present below the hydrocarbons. Water, as with all liquids, is compressible to a small degree. As the hydrocarbons are depleted, the reduction in pressure in the reservoir allows the water to expand slightly. Although this unit expansion is minute, if the aquifer is large enough this will translate into a large increase in volume, which will push up on the hydrocarbons, maintaining pressure.

With a water-drive reservoir the decline in reservoir pressure is very slight; in some cases the reservoir pressure may remain unchanged. The gas/oil ratio also remains stable. The oil rate will remain fairly stable until the water reaches the well. In time, the water cut will increase and the well will be watered out.

The water may be present in an aquifer (but rarely one replenished with surface water). This water gradually replaces the volume of oil and gas that is produced out of the well, given that the production rate is equivalent to the aquifer activity. That is, the aquifer is being replenished from some natural water influx. If the water begins to be produced along with the oil, the recovery rate may become uneconomical owing to the higher lifting and water disposal costs.

Water and Gas Injection

If the natural drives are insufficient, as they very often are, then the pressure can be artificially maintained by injecting water into the aquifer or gas into the gas cap.

Gravity Drainage

The force of gravity will cause the oil to move downward of the gas and upward of the water. If vertical permeability exists then recovery rates may be even better.

Gas and Gas Condensate Reservoirs

These occur if the reservoir conditions allow the hydrocarbons to exist as a gas. Retrieval is a matter of gas expansion. Recovery from a closed reservoir (i.e., no water drive) is very good, especially if bottom hole pressure is reduced to a minimum (usually done with compressors at the well head). Any produced liquids are light coloured to colourless, with a gravity higher than 45 API. Gas cycling is the process where dry gas is injected and produced along with condensed liquid.

Reservoir Fluids

Petroleum reservoir fluids are naturally occurring mixtures of gas and oil that exist at elevated pressures and temperatures. Reservoir fluid compositions typically include hundreds or thousands of hydrocarbons and a few non-hydrocarbons, like nitrogen, carbon dioxide and hydrogen sulphide. The physical properties of these mixtures depend primarily on composition, pressure, temperature and volume (PVT) conditions. Accurate data for the phase behaviour of these hydrocarbon mixtures is needed to improve oil recovery. Sometimes, experimental PVT data is available, but not many measurements are carried out for a given mixture; furthermore, it is expensive to investigate the full range of phase behaviour that can occur during a recovery process or a separation chain.

In the absence of PVT laboratory data, predictions of the reservoir fluid behaviour can be obtained by using a well-established equation of state to compute the phase behaviour and a correct representation of the composition of the reservoir fluid. Therefore, reservoir engineers must rely on characterization schemes and thermodynamic models to calculate the missing data.

The objective of this communication is to show how reservoir characterizations and PVT analyses can be carried out using VMGSim. A description of the new PVT Analysis unit operation, new to *VMGSim 9.0*, is included in this document; this unit operation is able to calculate the most common properties measured in PVT experiments and can be used to adjust characterization or thermodynamic model parameters to fit calculations to experimental data.

Reservoir Fluid Representation

The nature and composition of a reservoir fluid depends on the depositional environment of the formation from which the fluid is produced. Crude oil and natural gas are composed of many compounds with a wide range of molecular weights. Some estimates suggest that perhaps 3,000 organic compounds exist in a single reservoir fluid. The lighter and simpler compounds are produced as natural gas after surface separation, whereas the heavier and more complex compounds are obtained from crude oil at stock tank conditions.

Reservoir fluids fall into three broad categories; (i) aqueous solutions with dissolved salts, (ii) liquid hydrocarbons, and (iii) gases (hydrocarbon and non-hydrocarbon). In all cases their compositions depend upon their source, history, and present thermodynamic conditions. Their distribution within a given reservoir depends upon the thermodynamic conditions of the reservoir as well as the petrophysical properties of the rocks and the physical and chemical properties of the fluids themselves.

Fluid Distribution

The distribution of a particular set of reservoir fluids depends not only on the characteristics of the rock-fluid system now, but also the history of the fluids, and ultimately their source. A list of factors affecting fluid distribution would be manifold. However, the most important are:

- Depth: The difference in the density of the fluids results in their separation over time due to gravity (differential buoyancy).

- Fluid Composition: The composition of the reservoir fluid has an extremely important control on its pressure-volume-temperature properties, which define the relative volumes of each fluid in a reservoir. It also affects distribution through the wettability of the reservoir rocks.

- Reservoir Temperature: Exerts a major control on the relative volumes of each fluid in a reservoir.

- Fluid Pressure: Exerts a major control on the relative volumes of each fluid in a reservoir.

- Fluid Migration: Different fluids migrate in different ways depending on their density, viscosity, and the wettability of the rock. The mode of migration helps define the distribution of the fluids in the reservoir.

- Trap-Type: Clearly, the effectiveness of the hydrocarbon trap also has a control on fluid distribution (e.g., cap rocks may be permeable to gas but not to oil).

- Rock structure: The microstructure of the rock can preferentially accept some fluids and not others through the operation of wettability contrasts and capillary pressure. In addition, the common heterogeneity of rock properties results in preferential fluid distributions throughout the reservoir in all three spatial dimensions.

The fundamental forces that drive, stabilise, or limit fluid movement are:

- Gravity (e.g. causing separation of gas, oil and water in the reservoir column).

- Capillary (e.g. responsible for the retention of water in micro-porosity).

- Molecular diffusion (e.g. small scale flow acting to homogenise fluid compositions within a given phase).

- Thermal convection (convective movement of all mobile fluids, especially gases).

- Fluid pressure gradients (the major force operating during primary production).

Although each of these forces and factors vary from reservoir to reservoir, and between lithologies within a reservoir, certain forces are of seminal importance. For example, it is gravity that ensures, that when all three basic fluids types are present in an uncompartmentalised reservoir, the order of fluids with increasing depth is GAS:OIL:WATER, in exact analogy to a bottle of french dressing that has been left to settle.

Aqueous Fluids

Accumulations of hydrocarbons are invariably associated with aqueous fluids (formation waters), which may occur as extensive aquifers underlying or interdigitated with hydrocarbon bearing layers, but always occur within the hydrocarbon bearing layers as connate water. These fluids are commonly saline, with a wide range of compositions and concentrations; Table shows an example of a reservoir brine. Usually the most common dissolved salt is NaCl, but many others occur in varying smaller quantities. The specific gravity of pure water is defined as unity, and the specific gravity of formation waters increases with salinity at a rate of about 0.075 per 100 parts per thousand of dissolved solids. When SCAL measurements are made with brine, it is usual to make up a simulated formation brine to a recipe such as that given in table below, and then deaerate it prior to use.

Table: Composition of Draugen 6407/9-4 Formation Water.

Component	Concentration, g dm^{-3}
Pure water	Solvent
NaCl	34.70
$CaCl_2.6H_2O$	4.90
$MgCl_2.6H_2O$	2.70

KCl	0.40
NaHCO$_3$	0.40
SrCl$_2$.6H$_2$O	0.12
BaCl$_2$.6H$_2$O	0.06
Final pH = 7	

Why a connate water phase is invariably present in hydrocarbon bearing reservoir rock is easily explained. The reservoir rocks were initially fully or partially saturated with aqueous fluids before the migration of the oil from source rocks below them. The oil migrates upwards from the source rocks, driven by the differential buoyancy of the oil and the water. In this process most of the water swaps places with the oil since no fluids can escape from the cap rock above the reservoir. However, the water is not completely displaced as the initial reservoir rock is invariably water-wet, leaving the water-wet grains covered in a thin layer of water, with the remainder of the pore space full of oil. Water also remains in the microporosity where gravity segregation forces are insufficient to overcome the water-rock capillary forces.

The aqueous fluids, whether as connate water or in aquifers, commonly contain dissolved gases at reservoir temperatures and pressures. Different gases dissolve in aqueous fluids to different extents, and this gas solubility also varies with temperature and pressure. Table shows a selection of gases. If gas saturated water at reservoir pressure is subjected to lower pressures, the gas will be liberated, in exactly the same way that a lemonade bottle fizzes when opened. In reservoirs the dissolved gas is mainly methane (from 10 SCF/STB at 1000 psi to 35 SCF/STB at 10 000 psi for gas-water systems, and slightly less for water-oil systems). Higher salinity formation waters tend to contain less dissolved gas.

Table: Dissolution of Gases in Water (dissolved mole fraction) at 1 bar.

Gas	$10^4 \times$ Xg As @ 1 bar	
	25°C	55°C
Helium	0.06983	0.07179
Argon	0.2516	0.1760
Radon	1.675	0.8911
Hydrogen	0.1413	0.1313
Nitrogen	0.1173	0.08991
Oxygen	0.2298	0.0164
Carbon dioxide	6.111	3.235
Methane	0.2507	0.1684
Ethane	0.3345	0.01896
Ammonium	1876	1066
X$_{gas}$ = mole fraction of gas dissolved at 1 bar pressure, i.e.=1/H$_{gas}$.		

Aqueous fluids are relatively incompressible compared to oils, and extremely so compared to gases (2.5×10^{-6} to 5×10^{-6} per psi decreasing with increasing salinity). Consequently, if a unit volume of formation water with no dissolved gases at reservoir pressure conditions is transported to surface pressure condition, it will expand only slightly compared to the same initial volume of oil or gas. It should be noted that formation waters containing a significant proportion of dissolved gases are more compressible than those that are not gas saturated. These waters expand slightly more on being brought to the surface. However the reduction in temperature on being brought to the surface causes the formation water to shrink and there is also a certain shrinkage associated with the release of gas as pressure is lowered. The overall result is that brines experience a slight shrinkage (< 5%) on being brought from reservoir conditions to the surface.

Formation waters generally have densities that are greater than those of oils, and dynamic viscosities that are a little lower. The viscosity at high reservoir temperatures (>250oC) can be as low as 0.3 cP, rises to above 1 cP at ambient conditions, and increases with increasing salinity.

Table: densities and viscosities for a typical formation water and a refined oil.

Brine Component		Composition, g/l	
Pure water		Solvent	
NaCl		150.16	
$CaCl_2.6H_2O$		101.32	
$MgCl_2.6H_2O$		13.97	
Na_2SO_4		0.55	
$NaHCO_3$		0.21	
Fluid	Temperature, oC	Density, g/cm³	Dynamic Viscosity, cP
Brine	20	1.1250	1.509
Brine	25	1.1237	1.347
Brine	30	1.1208	1.219
Kerosene	20	0.7957	1.830
Kerosene	25	0.7923	1.661
Kerosene	30	0.7886	1.514

Phase Behavior of Hydrocarbon Systems

Figure shows the pressure versus volume per mole weight (specific volume) characteristics of a typical pure hydrocarbon (e.g. propane). Imagine in the following discussion that all changes occur isothermally (with no heat flowing either into or out of the fluid) and at the same temperature. Initially the component is in the liquid phase at 1000 psia, and has a volume of about 2 ft3 /lb.mol. (point A). Expansion of the system (A→B) results in large drops in pressure with small increases in specific volume, due to the small compressibility of liquids (liquid hydrocarbons as well as liquid formation waters have small compressibilities that are almost independent of pressure for the range of pressures encountered in hydrocarbon reservoirs). On further expansion, a pressure will be attained where the first tiny bubble of gas appears (point B). This is the bubble point or saturation pressure for a given temperature. Further expansion (B→C) now occurs at constant pressure with more and more of the liquid turning into the gas phase until no more fluid remains. The constant pressure at which this occurs is called the vapour pressure of the fluid at a given temperature. Point C represents the situation where the last tiny drop of liquid turns into gas, and is called the dew point. Further expansion now takes place in the vapour phase (C→D). The pistons in Figure demonstrate the changes in fluid phase schematically. It is worth noting that the process A®B®C®D described above during expansion (reducing the pressure on the piston) is perfectly reversible. If a system is in state D, then application of pressure to the fluid by applying pressure to the pistons will result in changes following the curve D→C→B→A.

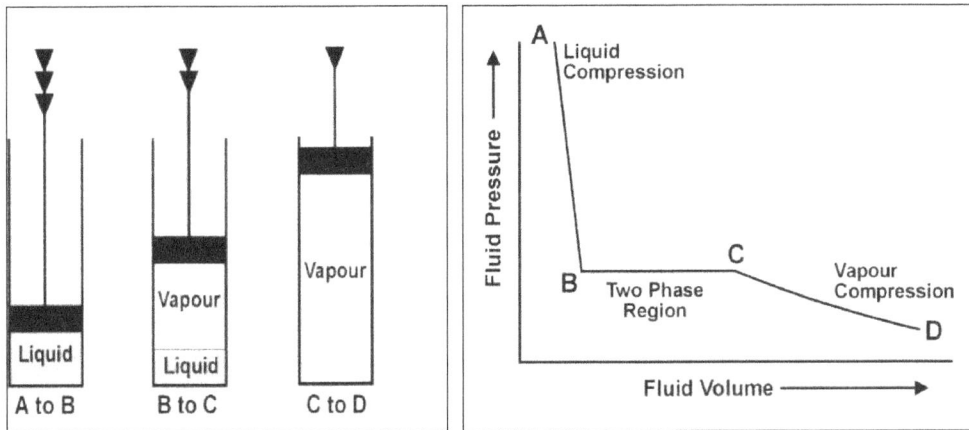

PV phase behaviour of a pure components.

We can examine the curve in Figure for a range of fluid temperatures. If this is done, the pressure-volume relationships obtained can be plotted on a pressure-volume diagram with the bubble point and dew point locus also included. Note that the bubble point and dew point curves join together at a point. This is the critical point. The region under the bubble point/dew point envelope is the region where the vapour phase and liquid phase can coexist, and hence have an interface (the surface of a liquid drop or of a vapour bubble). The region above this envelope represents the region where the vapour phase and liquid phase do not coexist.

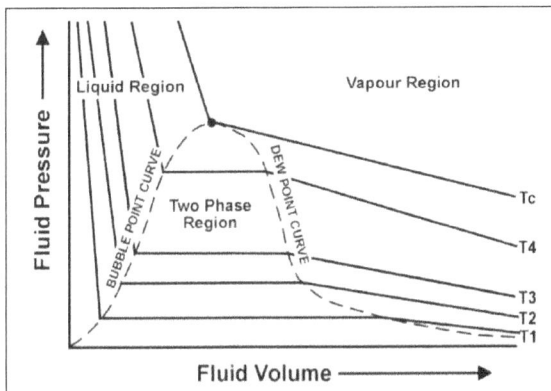

PV phase behaviour of a pure components.

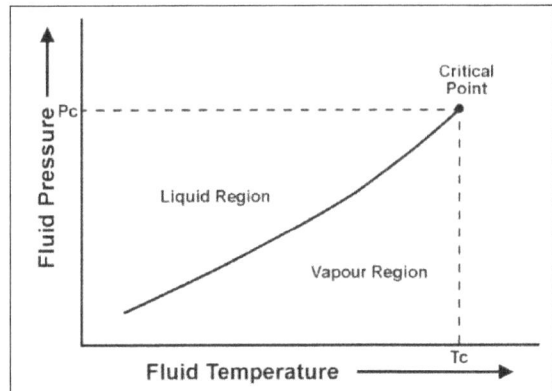

PT phase behaviour of a pure components.

Thus at any given constant low fluid pressure, reduction of fluid volume will involve the vapour condensing to a liquid via the two phase region, where both liquid and vapour coexist. But at a given constant high fluid pressure (higher than the critical point), a reduction of fluid volume will involve the vapour phase turning into a liquid phase without any fluid interface being generated (i.e. the vapour becomes denser and denser until it can be considered as a light liquid). Thus the critical point can also be viewed as the point at which the properties of the liquid and the gas become indistinguishable (i.e. the gas is so dense that it looks like a low density liquid and vice versa).

Suppose that we find the bubble points and dew points for a range of different temperatures, and plot the data on a graph of pressure against temperature. Figure shows such a plot. Note that the dew point and bubble points are always the same for a pure component, so they plot as a single line until the peak of figure is reached, which is the critical point.

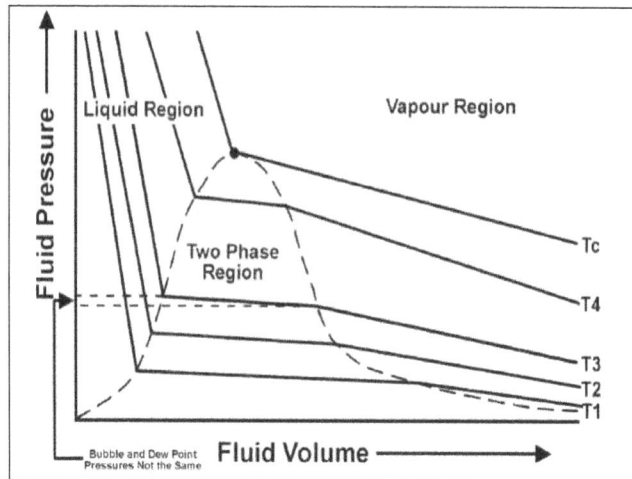

PV phase behaviour of a pure components.

The behaviour of a hydrocarbon fluid made up of many different hydrocarbon components shows slightly different behaviour. The initial expansion of the liquid is similar to that for the single component case.

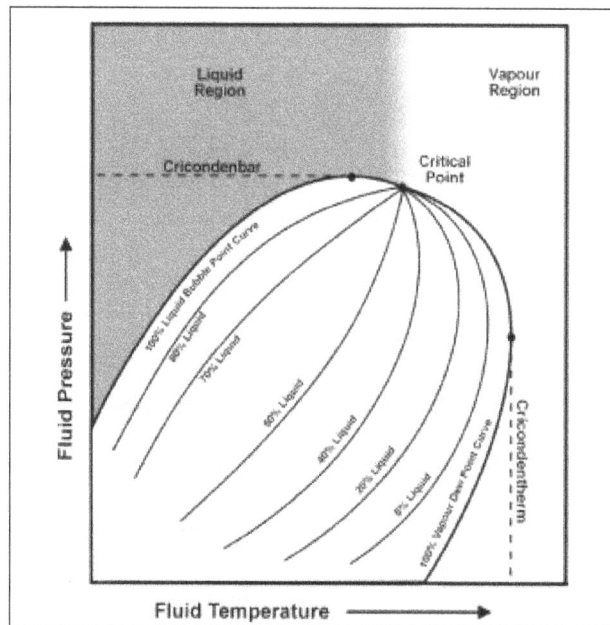

PT phase diagram for a multi components Fluid.

Once the bubble point is reached, further expansion does not occur at constant pressure but is accompanied by a decrease in pressure (vapour pressure) due to changes in the relative fractional amounts of liquid to gas for each hydrocarbon in the vaporising mixture. In this case the bubble points and dew points differ, and the resulting pressure-temperature plot is no longer a straight line but a phase envelope composed of the bubble point and dew point curves, which now meet at the critical point. There are also two other points on this diagram that are of interest. The cricondenbar, which defines the pressure above which the two phases cannot exist together whatever the temperature, and the cricondentherm, which defines the temperature above which the two phases cannot exist together whatever the pressure. A fluid that exists above the bubble point curve is classified as undersaturated as it contains no free gas, while a fluid at the bubble point curve or below it is classified as saturated, and contains free gas.

Figure shows the PT diagram for a reservoir fluid, together with a production path from the pressure and temperature existing in the reservoir to that existing in the separator at the surface. Note that the original fluid was an undersaturated liquid at reservoir conditions. On production the fluid pressure drops fast with some temperature reduction occurring as the fluid travels up the borehole. All reservoirs are predominantly isothermal because of their large thermal inertia. This results in the production path of all hydrocarbons initially undergoing a fluid pressure reduction. Figure shows that the ratio of vapour to liquid at separator conditions is approximately 55:45. If we analyse the PT characteristics of the separator gas and separator fluid separately then we would find that the separator pressuretemperature point representing the separator conditions falls on the dew point line of the separator gas PT diagram, and on the bubble point line of the separator oil PT diagram. This indicates that the shape of the PT diagram for various mixtures of hydrocarbon gases and liquids varies greatly. Clearly, therefore it is extremely important to understand the PT phase envelope as it can be used to classify and understand major hydrocarbon reservoirs.

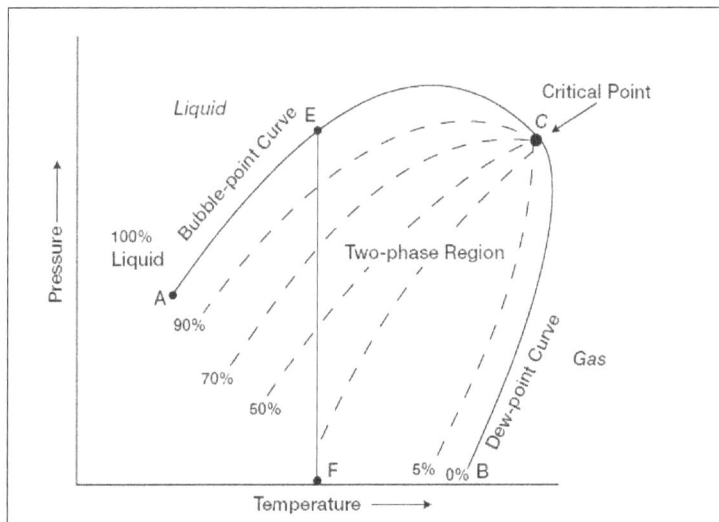

PT phase diagram for separator reservoir Fluid.

PVT Properties of Hydrocarbon Fluids

Cronquist Classification

Hydrocarbon reservoirs are usually classified into the following five main types, after Cronquist:

- Dry gas,
- Wet gas,
- Gas condensate,
- Volatile oil,
- Black oil.

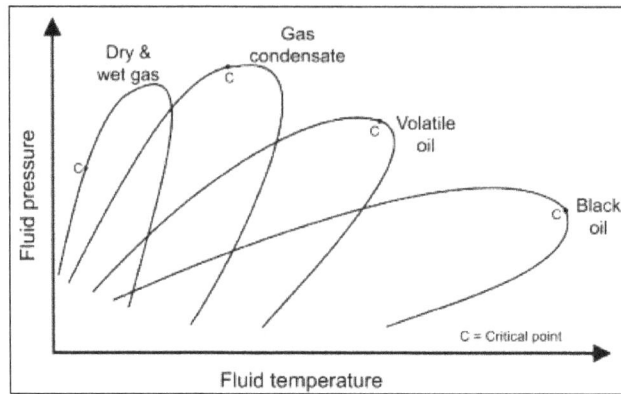

Each of these reservoirs can be understood in terms of its phase envelope.
The typical components of production from each of these reservoirs is shown in
table, and a schematic diagram of their PT phase envelopes is shown in figure.

Table: Typical Mol% compositions of fluids produced from cronquist reservoir types.

Component or Property	Dry Gas	Wet Gas	Gas Condensate	Volatile Oil	Black Oil
CO_2	0.10	1.41	2.37	1.82	0.02
N_2	2.07	0.25	0.31	0.24	0.34
C_1	86.12	92.46	73.19	57.60	34.62
C_2	5.91	3.18	7.80	7.35	4.11
C_3	3.58	1.01	3.55	4.21	1.01
iC_4	1.72	0.28	0.71	0.74	0.76
nC_4	-	0.24	1.45	2.07	0.49
iC_5	0.50	0.13	0.64	0.53	0.43
nC_5	-	0.08	0.68	0.95	0.21
C_6s	-	0.14	1.09	1.92	1.16
$C7+$	-	0.82	8.21	22.57	56.40
GOR (SCF/STB)	∞	69000	5965	1465	320
OGR (STB/MMSCF)	0	15	165	680	3125
API Specific Gravity, γ_{API} ,°API	-	65.0	48.5	36.7	23.6
C_{7+} Specific Gravity, g_o	-	0.750	0.816	0.864	0.920

Fundamental specific gravity g_o is equal to the density of the fluid divided by the density of pure water, and that for C_{7+} is for the bulked C_{7+} fraction. The API specific gravity γ_{API} is defined as; $\gamma_{API} = (141.5/g_o) - 131.5$.

Dry Gas Reservoirs

A typical dry gas reservoir is shown in figure. The reservoir temperature is well above the cricondentherm. During production the fluids are reduced in temperature and pressure. The temperature-pressure path followed during production does not penetrate the phase envelope, resulting in the production of gas at the surface with no associated liquid phase. Clearly, it would be possible to produce some liquids if the pressure is maintained at a higher level. In practice, the stock tank pressures are usually high enough for some liquids to be produced. Note the lack of C_{5+} components, and the predominance of methane in the dry gas in table.

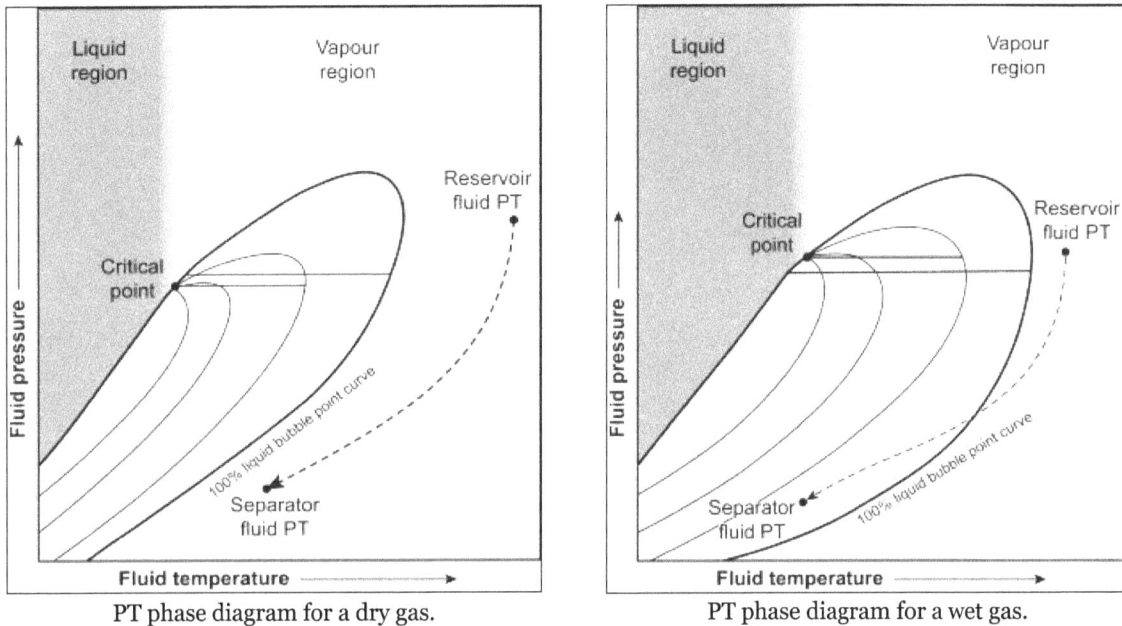

PT phase diagram for a dry gas.

PT phase diagram for a wet gas.

Wet Gas Reservoirs

A typical wet gas reservoir is shown in figure .The reservoir temperature is just above the cricondentherm. During production the fluids are reduced in temperature and pressure. The temperature-pressure path followed during production just penetrates the phase envelope, resulting in the production of gas at the surface with a small associated liquid phase. Note the presence of small amounts of C_{5+} components, and the continuing predominance of methane in the wet gas in table. The GOR (gas-oil ratio) has fallen as some liquid is being produced. However, this liquid usually amounts to less than about 15 STB/MMSCF. Note also the small specific gravity for C_{7+} components (0.750), indicating that the majority of the C_{7+} fraction is made up of the lighter C_{7+} hydrocarbons.

Gas Condensate Reservoirs

A typical gas condensate reservoir is shown in figure The reservoir temperature is such that it falls between the temperature of the critical point and the cricondentherm. The production path then has a complex history. Initially, the fluids are in an indeterminate vapour phase, and the vapour expands as the pressure and temperature drop. This occurs until the dewpoint line is reached, whereupon increasing amounts of liquids are condensed from the vapour phase. If the pressures and temperatures reduce further, the condensed liquid may reevaporate, although sufficiently low pressures and temperatures may not be available for this to happen. If this occurs, the process is called isothermal retrograde condensation. Isobaric retrograde condensation also exists as a scientific phenomenon, but does not occur in the predominantly isothermal conditions of hydrocarbon reservoirs. Thus, in gas condensate reservoirs, the oil produced at the surface results from a vapour existing in the reservoir. Note the increase in the C_{7+} components and the continued importance of methane. The GOR has decreased significantly, the OGR has increased, and the specific gravity of the C_{7+} components is increasing, indicating that greater fractions of denser hydrocarbons are present in the C_{7+} fraction.

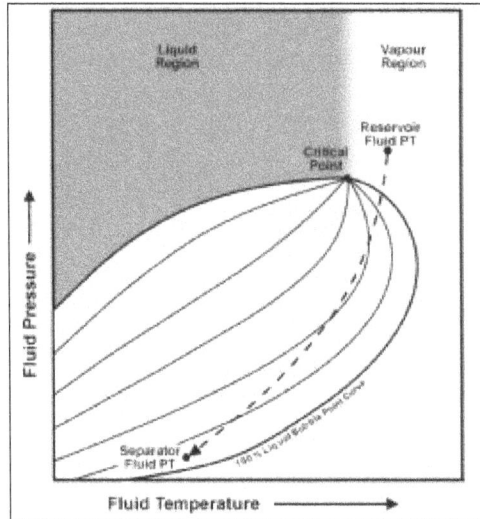

PT phase diagram for a gas condensate.

Volatile Oil Reservoirs

A typical volatile oil reservoir is shown in figure. The reservoir PT conditions place it inside the phase envelope, with a liquid oil phase existing in equilibrium with a vapour phase having gas condensate compositions. The production path results in small amounts of further condensation, and re-evaporation can occur again, but should be avoided as much as possible by keeping the stock tank pressure as high as possible. Reference to Table shows that the fraction of gases is reduced, and the fraction of denser liquid hydrocarbon liquids is increased, compared with the previously discussed reservoir types. Changes in the GOR, OGR and specific gravities are in agreement with the general trend.

Black Oil Reservoirs

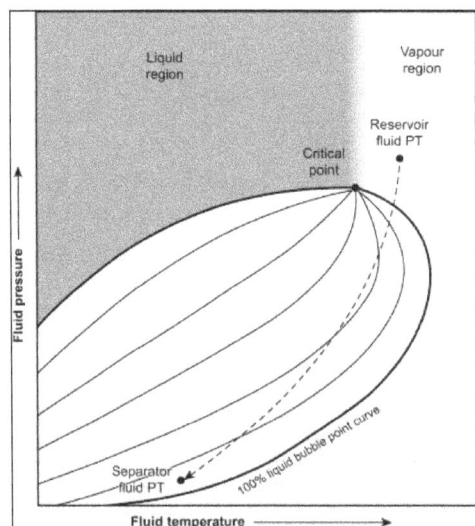

PT phase diagram for a volatile.

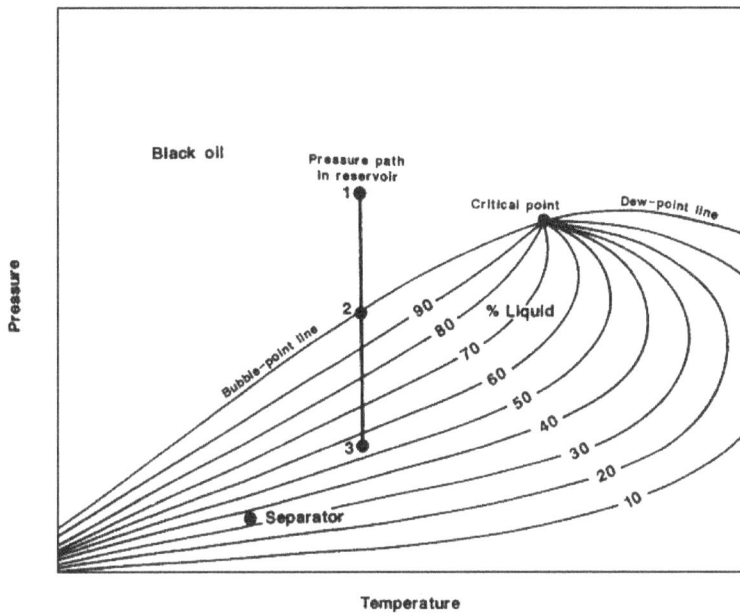

PT phase diagram for a black oil.

A typical gas condensate reservoir is shown in figure .The reservoir temperature is much lower than the temperature of the critical point of the system, and at pressures above the cricondenbar. Thus, the hydrocarbon in the reservoir exists as a liquid at depth. The production path first involves a reduction in pressure with only small amounts of expansion in the liquid phase. Once the bubble point line is reached, gas begins to come out of solution and continues to do so until the stock tank is reached. The composition of this gas changes very little along the production path, is relatively lean, and is not usually of economic importance when produced. Table shows a produced hydrocarbon fluid that is now dominated by heavy hydrocarbon liquids, with most of the produced gas present as methane. The GOR, OGR and specific gravities mirror the fluid composition.

References

- Petroleum-geology-8255: petropedia.com, Retrieved 24 July, 2019

- Hyne N.J. (2001). Nontechnical Guide to Petroleum Geology, Exploration, Drilling, and Production. PennWell Books. p. 164. ISBN 9780878148233

- Reservoir-rock, engineering: sciencedirect.com, Retrieved 10 May, 2019

- Types-of-caprocks-in-petroleum-system: geologylearn.blogspot.com, Retrieved 26 March, 2019

- "Rise and Fall of the North Sea". Schlumberger Excellence in Educational Development. Archived from the original on November 22, 2005. Retrieved January 30, 2006

- Formation-petroleum-reservoirs, reservoir-fundamentals: appliedpetroleumreservoirengineering.com, Retrieved 11 April, 2019

- Gluyas, J; Swarbrick, R (2004). Petroleum Geoscience. Blackwell Publishing. p. 148. ISBN 978-0-632-03767-4

- Reservoir-Fluid-Characterization: virtualmaterials.com, Retrieved 24 August, 2019

Chapter 3

Petroleum Exploration

The search for petroleum and natural gas within the Earth by using methods and techniques of petroleum geology is called petroleum exploration. The diverse applications of petroleum exploration as well as the techniques used within it, such as well logging, have been thoroughly discussed in this chapter.

Exploration is a method used for searching potentially viable oil and gas sources through geological surveys and drilling exploration wells to identify areas of potential interest. Petroleum exploration also known as or oil and gas exploration or Hydrocarbon exploration, is the the procedure for deposition of hydrocarbons, particularly petroleum and natural gas in the Earth.

The role of exploration is to provide the information required to exploit the best opportunities presented in the choice of areas, and to manage research operations on the acquired blocks.

An oil company may work for several years on a prospective area before an exploration well is spudded and during this period the geological history of the area is studied and the likelihood of hydrocarbons being present quantified.

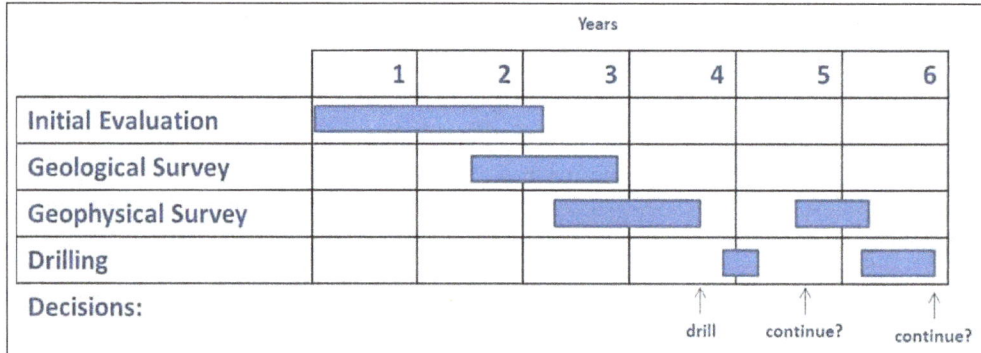

	Years					
	1	2	3	4	5	6
Initial Evaluation	███	███				
Geological Survey		███				
Geophysical Survey			███		███	
Drilling				▮		███
Decisions:				↑ drill	↑ continue?	↑ continue?

Stages of a typical exploration program.

Exploration is responsible for handling the risk intrinsic in this activity, and this is generally achieved by selection of a range of options in probabilistic and economic terms. Indeed, exploration is a risk activity and the management of exploration assets and associated operations is a major task for oil companies.

The risk cannot be eliminated entirely but can be controlled and reduced adopting appropriate workflow, conceptual and technological innovations.

When it's been decided to start up with an exploration project in a basin or in a larger area containing several basins, the quantity and quality of available data must be acquired and evaluated – geological data, type of reserves, production of existing fields (if any), etc.

Basin assessment/evaluation is the first step to undertake the study of the area under interest.

Technological development has provided oil companies with Basin Modeling – which is a numerical simulations that allows the temporal reconstruction of the history of a sedimentary basin and the associated evolution of the processes related to the formation of petroleum accumulations.

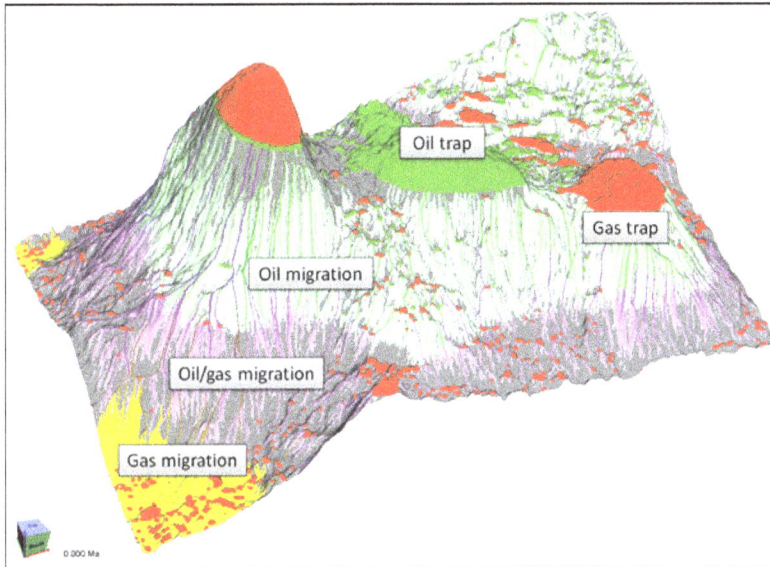

Basin modeling – Petroleum system.

On the basis of data and evidences collected from the preliminary studies, the company management, in the light of the possibilities and the probabilities of a discovery based on G&G data, aside from considerations of an economic nature, may decide to move to the following stage, which is the acquisition (through direct negotiations or by taking part in bids, etc.) of the legal right to perform prospecting in the selected area/block.

The owner of the mining right is normally the State, with which the oil company stipulates a contract establishing the contracting parties' rights.

Production Sharing Contracts and service contracts are frequently adopted nowadays. The sequence of activities covered by an exploration permit is fairly uniform, and include:

- The creation of a database.

- The analysis of available data.

- The programming of mapping and geological and photo-geological surveys.

- Seismic surveys and interpretation of seismic data.

- The choice of well locations, drilling.

- The analysis of results and the decision as to whether or not to proceed with the application for a lease or to release the area after fulfilling obligations.

Goal of exploration is to identify and locate a prospect, to quantify the volume of hydrocarbon which might be contained in the potential reservoirs and to evaluate the risk inherent the project itself.

A prospect is a viable target evidenced by geological and geophysical indications that is recommended for drilling an exploration well.

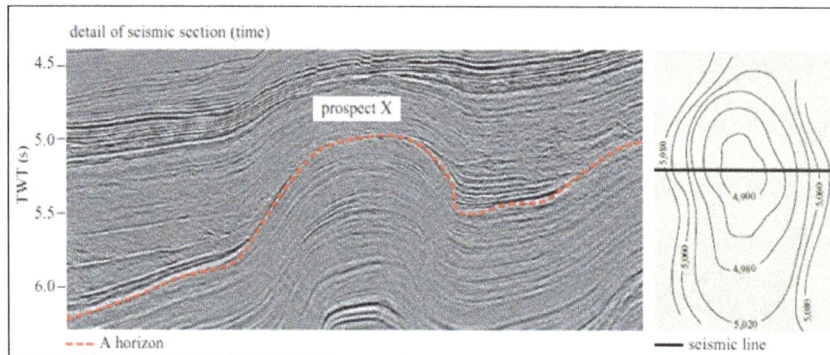

A detail of seismic section (time) and A horizon
depth map (meters) referred to prospect X.

The prospects identified must be technically practicable and meet the market conditions to guarantee a financial return on investments. The results obtained by drilling the exploratory wells indicate whether the initial geological hypotheses are correct or whether variations are found.

All this will allow the fine-tuning of the economic analysis of the project possibly turning hypothetical reserves into proved ones. Where profitability does not meet the standards of the company, it leads to the termination of further investments.

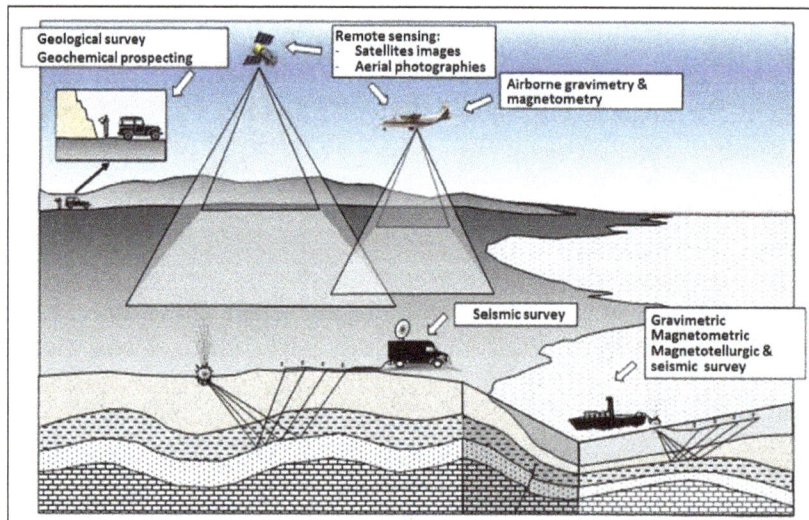

The main petroleum exploration techniques.

Geological Mapping and Prospecting

Geological mapping and prospecting are valuable techniques in an petroleum exploration.

Geological mapping It is basically a technique which allows a graphical presentation of geological observations and interpretations. Geological prospecting make use of geological disciplines such as petrography, stratigraphy, sedimentology, structural geology, geochemistry. Such disciplines are used to achieve different targets but it must be stressed that their integration is fundamental to depict a picture of reality.

Geophysical Methods

Geophysical methods allow to study the physical properties of the subsurface rocks and they can be used in different phases of the exploration in order to collect different types of information.

Geophysical methods such as gravimetric, magnetometric, magnetotelluric, seismic are often combined to obtain more accurate and reliable results.

Gravimetric Prospecting

- Gravimetric prospecting is a geophysical technique which is able to identify anomalies in the gravity acceleration generated by contrasts in density among bodies in the subsurface.

- Gravimetric prospecting is used to reconstruct of the main structural elements of sedimentary basins such as:
  - Extension, thickness, salt domes, intrusive plutons and dislocations or fault lines.

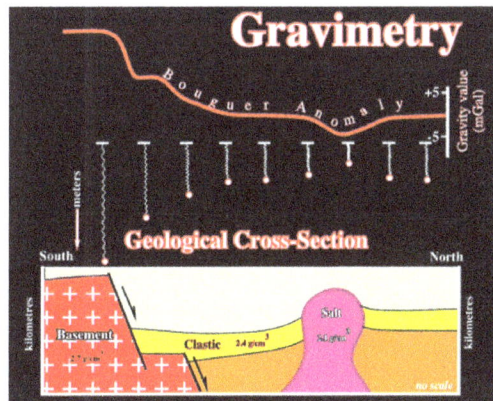

Magnetometric Prospecting

- This method involves measuring local anomalies in the Earth's magnetic fields.

- The method enable acquisition of data on structural characteristics and depth of the susceptive basement and therefore, indirectly, on the thickness of sedimentary overburden and identifies the presence, depth and extension of volcanic or plutonic masses within the sedimentary sequences.

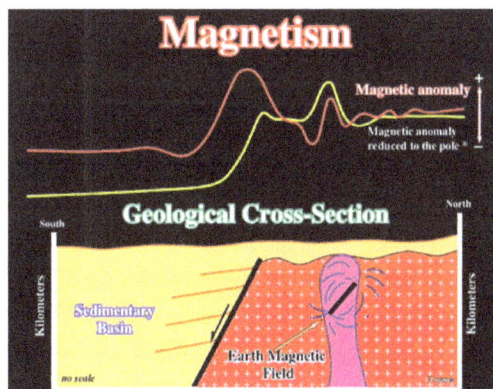

Seismic Prospecting

- Seismic prospecting has become the most valuable technique to reduce exploration risk of being unsuccessful in locating a prospect.

- The technique is based on determinations of the time interval that elapses between the initiation of a seismic wave at a selected shop point and the arrival of reflected or refracted impulses at one or more seismic detectors.

- The phase of seismic data acquisition is followed by the seismic data processing phase (aimed to the alteration of seismic data to suppress noise, enhance signal and migrate seismic events to the appropriate location in space) than by the interpretation of the generated subsurface image.

Well Logging

Logging is a continuous recording of the physical properties of rocks in the well with respect to depth. These physical properties are porosity, resistivity, density, conductivity, saturation etc. Logging was started with simple electric logs measuring the electric conductivity of rocks, but it is now a technically advanced and sophisticated method. Logging plays a crucial role in exploration and exploitation of hydrocarbons.

Well logging in oil industry has its own meaning; log means "record against depth of any of the characteristics of the rock formations traversed by a measuring apparatus in the well bore". The value of the measurement is plotted continuously against depth in the well. For example, the resistivity log is a continuous plot of a formation resistivity from the bottom of the well to the top.

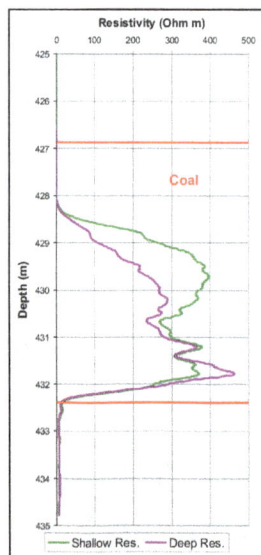

A well log.

The most appropriate name for this continuous depth related record is a wire- line geophysical well log, conveniently shortened to well log or log.

Generally, we go for geophysical methods for identifying the structures present beneath the earth subsurface over a few kilometres. One of the geophysical methods, that is logging is used to study the physical parameters of the formations inch by inch. Whenever an interesting zone is encountered we go for coring. This is the direct analysis of the formations. But it is not possible to take core analysis for entire well. So logging will give adequate data of entire well in indirect analysis.

Petro Physical Parameters

Porosity: Porosity is defined as a measure of the pore spaces present in a rock. These pore spaces within the reservoir rocks could be filled with gas, oil or water. High porosity values indicate high capacities of the reservoir rocks to contain these fluids, while low porosity values indicate the opposite. Porosity is calculated as the pore volume of the rock divided by its bulk volume. Expressed in terms of symbols, is represented as:

$$\phi = V_p / V_B$$

$$\phi - porosity$$

Where,

$$V_p - pore\,volume$$

$$V_B - vulk\,volume$$

Pore volume is the total volume of pore spaces in the rock, and bulk volume is physical volume of the rock, which includes the pore spaces and matrix materials (sand and shale, etc.) that compose the rock. Porosity has no units. It is expressed in percentages. Two types of porosities can exist in a rock. These are termed as:

- Primary porosity: Primary porosity is described as the porosity of the rock that formed at the time of its deposition.

- Secondary porosity: Secondary porosity develops after deposition of the rock. Secondary porosity includes vugular spaces in carbonate rocks created by the chemical process of leaching, or fracture spaces formed in fractured reservoirs. Porosity is further classified as:

  ◦ Total porosity is defined as the ratio of the entire pore space in a rock to its bulk volume.

  ◦ Effective porosity is the ratio of interconnected pore space to the bulk volume of the rock.

Rock porosity data are obtained by direct or indirect measurements. Laboratory measurements of porosity data on core samples are examples of direct methods. Determinations of porosity data from well log data are considered indirect methods.

Factors effecting porosity:

- Grain Shape
- Grain size
- Grain Packing
- Distribution Of Grains.

Permeability

Permeability is a measure of the ability of a porous medium, to transmit fluids through its interconnected pore spaces. Permeability is three types:

- Effective Permeability: When the pore spaces in the porous medium are occupied by more than one fluid, the permeability measured is the Effective permeability of the porous medium to that particular fluid. For instance, the effective permeability of a porous medium to oil is the permeability to oil when other fluids, including oil, occupy the pore spaces.

- Relative Permeability: The ratio of effective permeability to absolute permeability of a porous medium.

- Absolute permeability: If the porous medium is completely saturated (100% saturated) with a single fluid, the permeability measured is the absolute permeability. It is an intrinsic property of the porous medium, and the magnitude of absolute permeability is independent of the type of fluid in the pore spaces.

Wire-line Geophysical Logs

Wireline geophysical well logs (Open-hole log) are recorded when the drilling tools are no longer in the hole. Wireline logs are made using highly specialized equipment entirely separate from that used for drilling. Onshore, a motorized logging truck is used which brings its array of surface recorders, computers and logging drum and cable to the drill site. Offshore, the same equipment is installed in a small cabin left permanently on the rig. Both truck and cabin use a variety of inter-changeable logging tools, which are lowered into the well on the logging cable as shown in the below.

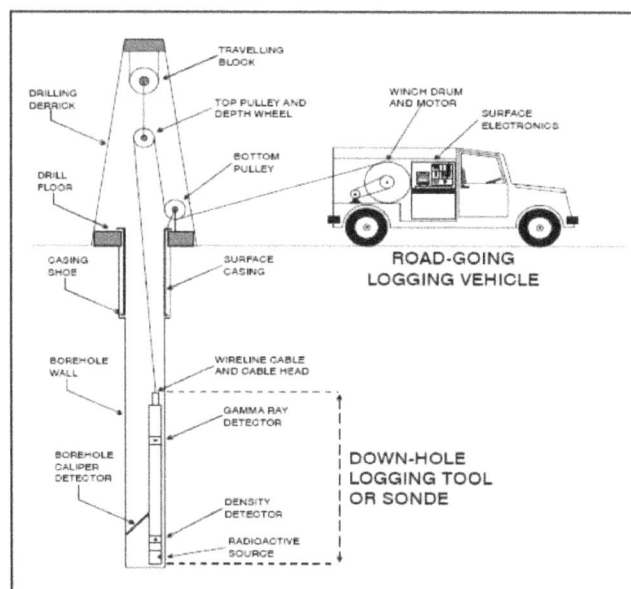

Modern Wire line logging set up.

Most modern logs are recorded digitally. The sampling rate will normally be once every 15 cm (6inch). At typical logging speeds, data transmission rates will vary from 0.05 kilobits per second for simpler logs to over 200 kilobits per second for the new complex logs. The huge amount of data representing each logging run is fed into the computer of the surface unit.

To run wireline logs, the hole is cleaned and stabilized and the drilling equipment extracted. The first logging tool is then attached to the logging cable (wireline) and lowered into the hole to its maximum drilled depth. Most logs are run while pulling the tool up from the bottom of the hole.

The cable attached to the tool acts both as a support for the tool and as a canal for the data transmission. The outside consists of galvanized cell, while the electrical conductors are insulated in the interior the cable is wound around a motorized drum on to which it is guided manually during logging.

The drum will pull the cable at speeds of between 300m/h (1000ft/h) and 1800m/h (6000ft/h). Logging cables have magnetic markets set a regular interval along their length and depths are checked mechanically, but apparent depths must be corrected for the cable tension and elasticity.

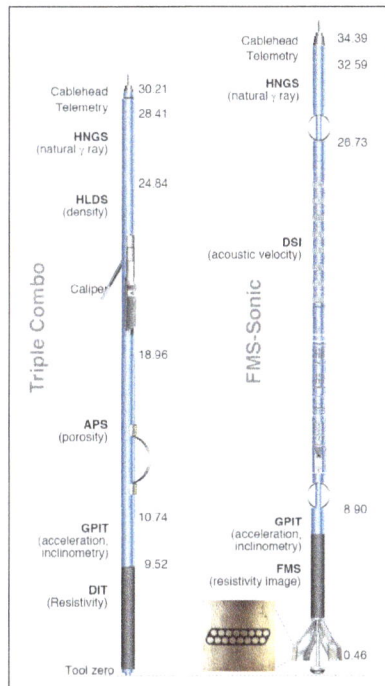

Typical Modern Combinational logging tools.

Modern logging tools are multifunction they may be up to 28m in length, but still have an overall diameter of only 3-4 in. For example 3⅜ in diameter, is 55.5 ft (16.9m) long and gives a simultaneous measurement of gamma ray or calliper, SP, deep resistivity, shallow resistivity and sonic velocity. The complexity of this tool requires the use of surface computer, not only to record but also to memorize and to depth-match the various readings.

Despite the use of the combined tools, the recording of a full set of logs still requires several different tool descents. While a quick, shallow logging job may only take 3hours, a deephole, full set may take 2-3 days, each tool taking perhaps 4-5hours to complete.

Borehole Environment

Borehole environment provides the limitations for tool design and operations. A true vertical hole is rarely encountered, and generally the deviation of the borehole is between 0° and 5°onshore. The temperature at full depth ranges between 100 °F and 300° F. The salinity of the drilling mud ranges between 3,000 and 2,00,000 ppm of NaCl. This salinity, coupled with the fact that the well bore is generally over pressured, causes invasion of a porous and permeable formation by the drilling fluid. In the permeable zone, due to the imbalance in hydrostatic pressure, the mud begins to enter the formation but is normally rapidly stopped by the build-up of a mud cake of the clay particles in the drilling fluid.

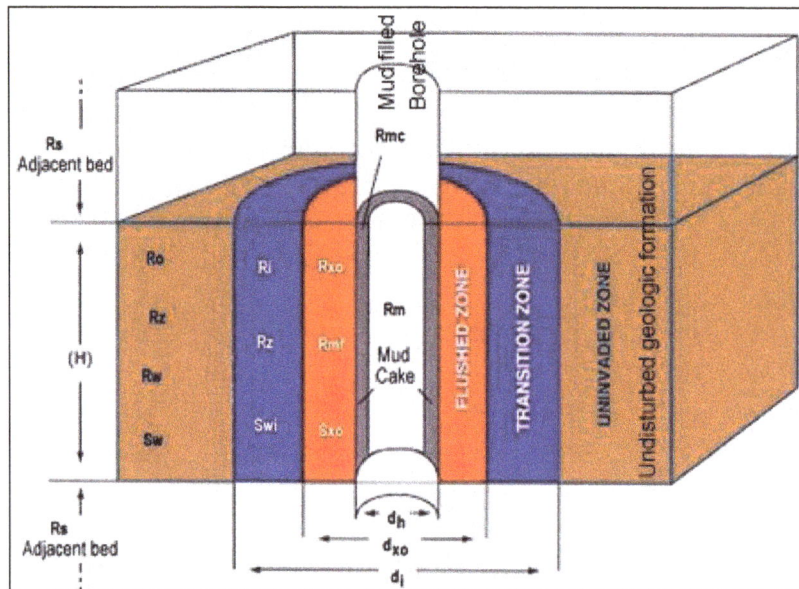

Bore Hole Environment.

Where,

- dh – hole diameter.

- d_I – Diameter of invaded zone (inner boundary; flushed zone).

- d_j – Diameter of invaded zone (outer boundary; invaded zone).

- D_{rJ} – Radius of invaded zone (outer boundary).

- h_{mc} –Thickness of mud cake.

- R_m – Resistivity of the drilling mud.

- R_{mc} – Resistivity of the mud cake.

- R_{mf} – Resistivity of mud filtrate Rs – Resistivity of shale.

- R_t – Resistivity of uninvaded zone (true resistivity) – Resistivity of formation water.

- R_{xo} – Resistivity of flushed zone.

- S_t – Water saturation of uninvaded zone.

- S_{xo} – Water saturation flushed zone.

Invasion

A Pressure induced migration of drilling mud fluids (mud filtrate) in to permeable formations. Mud cake is deposited on the borehole wall. Partially sealing the formation and slowing further invasion.

Very close to the borehole most of the original formation water and some of the hydrocarbons may be flushed away by the filtrate. This zone is referred to as the flushed zone (Invaded zone).

Further out from the borehole, the displacement of the formation fluids by the mud filtrate becomes lesser and lesser, resulting in a transition from mud filtrate saturation to original formation. This zone is referred to as the transition zone. The undisturbed formation beyond the transition zone is referred to as the non-invaded, virgin, or uncontaminated zone.

Log Representation

A standard API (American Petroleum Institute) log format exists. The overall log width is 8.25 in (21cm), with three tracks of 2.5 in (6.4cm), track 1 and 2 being separated by a column of 0.75 in (1.9cm) in which the depths are printed. There are various combination of grids. Track one is always linear, with ten standard divisions of 0.25in (0.64 cm). Depth track is presented after Track1. Track 2 and 3 may have a 4-cycle logarithmic scale, a linear scale of 20 standard divisions, or a hybrid of logarithmic scale in track 2 and linear scale in track 3.

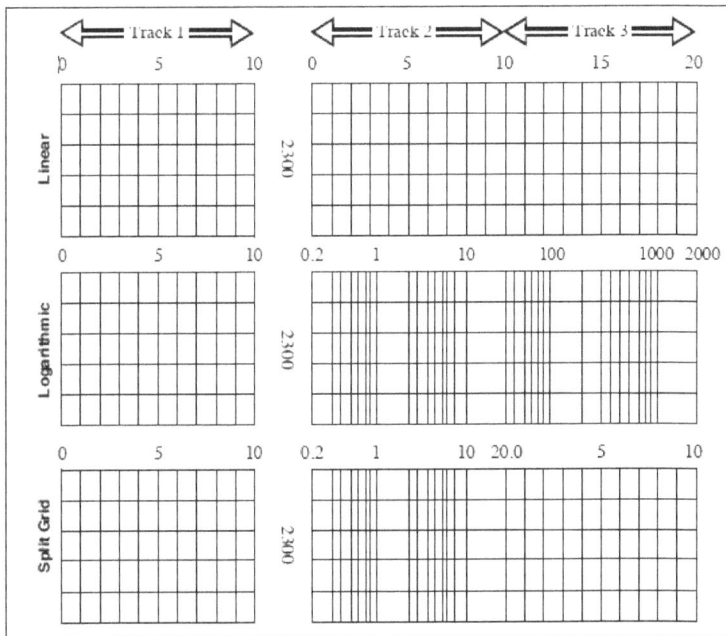

Three typical API log formats.

On the old analog logging system, the choice of vertical or depth scales was limited to two of 1:1000, 1:500, 1:200 , 1:100 , 1:40 and 1:20. From these, the most frequent scale combinations were 1:500(1cm = 5m) for correlation logs and 1:200(1cm=2m) for detailed reservoir presentation.

One of the final aspects of log grid is for markers which indicate real time during logging. The markers can be dashed or ticks or spikes on the grid, time markers allow a direct control of logging speed and indirectly on log quality.

Log Header.

Every log is preceded by a comprehensive log heading. It covers all aspects which allow the proper interpretation of the log and identification of well, rig, logger and logging unit. The log heading is an example, each company having its own format many surface geophysical methods such as seismic, gravity magnetic, magneto-telluric, geological field survey etc. are all indirect methods. They indicate the structural and stratigraphic location, where hydrocarbon accumulations might occur. They give no evidence, whether there is a significant accumulation at a depth of few miles below the surface of earth and their quantitative estimation. The only method that is available for answering these questions is the exploratory wells. Well log data provide all these parameters in-situ required to estimate the hydrocarbon reserves and their producibility. A large number of physical rock properties can now be measured by using different down hole tools and surface recording system also called logging unit or truck. They include resistivity, bulk mass density, interval transit time (just reciprocal of sonic velocity), spontaneous potential, natural radioactivity and hydrogen content of the rock.

Drilling Rig

A drilling rig is an integrated system that drills wells, such as oil or water wells, in the earth's subsurface. Drilling rigs can be massive structures housing equipment used to drill water wells, oil wells, or natural gas extraction wells, or they can be small enough to be moved manually by one person and such are called augers. Drilling rigs can sample subsurface mineral deposits, test rock, soil and groundwater physical properties, and also can be used to install sub-surface fabrications, such as underground utilities, instrumentation, tunnels or wells. Drilling rigs can be mobile equipment mounted on trucks, tracks or trailers, or more permanent land or marine-based structures (such as oil platforms, commonly called 'offshore oil rigs' even if they don't contain a drilling rig). The term "rig" therefore generally refers to the complex equipment that is used to penetrate the surface of the Earth's crust.

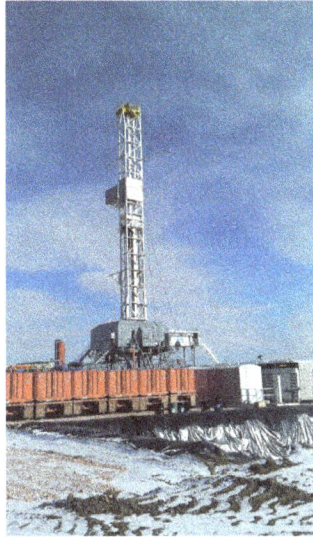

Drilling the Bakken Formation in the Williston Basin.

Small to medium-sized drilling rigs are mobile, such as those used in mineral exploration drilling, blast-hole, water wells and environmental investigations. Larger rigs are capable of drilling through thousands of metres of the Earth's crust, using large "mud pumps" to circulate drilling mud (slurry) through the drill bit and up the casing annulus, for cooling and removing the "cuttings" while a well is drilled. Hoists in the rig can lift hundreds of tons of pipe. Other equipment can force acid or sand into reservoirs to facilitate extraction of the oil or natural gas; and in remote locations there can be permanent living accommodation and catering for crews (which may be more than a hundred). Marine rigs may operate thousands of miles distant from the supply base with infrequent crew rotation or cycle.

Large hole drilling rig for blast-hole drilling.

Petroleum Drilling Industry

Oil and natural gas drilling rigs are used not only to identify geologic reservoirs but also to create holes that allow the extraction of oil or natural gas from those reservoirs. Primarily in onshore oil and gas fields once a well has been drilled, the drilling rig will be moved off of the well and a service rig (a smaller rig) that is purpose-built for completions will be moved on to the well to get the well on line. This frees up the drilling rig to drill another hole and streamlines the operation as well as allowing for specialization of certain services, i.e. completions vs. drilling.

Mining Drilling Industry

Mining drilling rigs are used for two main purposes, exploration drilling which aims to identify the location and quality of a mineral, and production drilling, used in the production-cycle for mining. Drilling rigs used for rock blasting for surface mines vary in size dependent on the size of the hole desired, and is typically classified into smaller pre-split and larger production holes. Underground mining (hard rock) uses a variety of drill rigs dependent on the desired purpose, such as production, bolting, cabling, and tunnelling.

Mobile Drilling Rigs

Mobile drilling rig mounted on a truck.

In early oil exploration, drilling rigs were semi-permanent in nature and the derricks were often built on site and left in place after the completion of the well. In more recent times drilling rigs are expensive custom-built machines that can be moved from well to well. Some light duty drilling rigs are like a mobile crane and are more usually used to drill water wells. Larger land rigs must be broken apart into sections and loads to move to a new place, a process which can often take weeks.

Small mobile drilling rigs are also used to drill or bore piles. Rigs can range from 100 ton continuous flight auger (CFA) rigs to small air powered rigs used to drill holes in quarries, etc. These rigs use the same technology and equipment as the oil drilling rigs, just on a smaller scale.

The drilling mechanisms outlined below differ mechanically in terms of the machinery used, but also in terms of the method by which drill cuttings are removed from the cutting face of the drill and returned to surface.

Drilling Rig Classification

There are many types and designs of drilling rigs, with many drilling rigs capable of switching or combining different drilling technologies as needed. Drilling rigs can be described using any of the following attributes:

Power Used

- Mechanical: The rig uses torque converters, clutches, and transmissions powered by its own engines, often diesel.

- Electric: The major items of machinery are driven by electric motors, usually with power generated on-site using internal combustion engines.

- Hydraulic: The rig primarily uses hydraulic power.

- Pneumatic: The rig is primarily powered by pressurized air.

- Steam: The rig uses steam-powered engines and pumps (obsolete after middle of 20th century).

Pipe Used

- Cable: A braided hemp or wire rope is used to raise and drop the drill bit.

- Conventional: Uses metal or plastic drill pipe of varying types.

- Coil tubing: Uses a single flexible tube of sufficient length, stored on a drum up to five metres in diameter, and a downhole drilling motor.

- Chain: A chain is used to raise and drop the drill bit in some hydraulic rigs.

Height

Rigs are differentiated by height based on how many connected joints of drill pipe they are able to "stand" in the derrick when it is out of the hole. Typically this is done when changing a drill bit or when "logging" the well. A single joint of pipe is typically some 30 feet long.

- Single: Can hold only single drill pipes. The presence or absence of vertical pipe racking "fingers" varies from rig to rig.

- Double: Derrick can hold two connected drill pipes, called a "double stand" or simply a "double".

- Triple: Derrick can hold three connected drill pipes: a "triple stand" or "thribble".

- Quadri: Can hold four connected drill pipes, called a "quadri stand" or "fourble". Derricks of this size are uncommon.

Method of Rotation or Drilling Method

- No-rotation includes direct push rigs and most service rigs.

- Rotary table: Rotation is achieved by turning a square or hexagonal pipe (the "Kelly") in a rotary table at drill floor level.

- Top drive: Rotation and circulation is done at the top of the drill string, on a motor that moves in a track along the derrick. This is the design of the most modern rigs.

- Sonic: Uses primarily vibratory energy to advance the drill string.

- Hammer: Uses rotation and percussive force.

Position of Derrick

- Conventional: Derrick is vertical.

- Slant: Derrick is slanted at a 45 degree angle to facilitate horizontal drilling.

Directional Drilling (DD)

Directional drilling is done with a mud motor on the Bottom hole assembly (BHA). The direction is controlled by wireless controller to drill the hole in any way that the driller requires. This is the future of drilling.

Drilling Techniques

There are a variety of drilling techniques which can be used to drill a borehole into the ground. Each has its advantages and disadvantages, in terms of the depth to what it can drill, the type of sample returned, the costs involved and penetration rates achieved. Some types included are Rotary cut, Rotary Abrasive, Rotary Reverse, Cable Tooling, and Sonic Drilling.

Auger Drilling

Auger drilling is done with a helical screw which is driven into the ground with rotation; the earth is lifted up the borehole by the blade of the screw. Hollow stem auger drilling is used for softer ground such as swamps where the hole will not stay open by itself for environmental drilling, geotechnical drilling, soil engineering and geochemistry reconnaissance work in exploration for mineral deposits. Solid flight augers/bucket augers are used in harder ground construction drilling. In some cases, mine shafts are dug with auger drills. Small augers can be mounted on the back of a utility truck, with large augers used for sinking piles for bridge foundations.

Auger drilling is restricted to soil, soft unconsolidated formations, or weak weathered rock. It is cheap and fast.

Percussion Rotary Air Blast Drilling (RAB)

RAB drilling is used most frequently in the mineral exploration industry. (This tool is also known as a Down-the-hole drill.) The drill uses a pneumatic reciprocating piston-driven "hammer" to energetically drive a heavy drill bit into the rock. The drill bit is hollow, solid steel and has ~20 mm thick tungsten rods protruding from the steel matrix as "buttons". The tungsten buttons are the cutting face of the bit.

The cuttings are blown up the outside of the rods and collected at surface. Air or a combination of air and foam lift the cuttings.

RAB drilling is used primarily for mineral exploration, water bore drilling and blast-hole drilling in mines, as well as for other applications such as engineering, etc. RAB produces lower quality samples because the cuttings are blown up the outside of the rods and can be contaminated from contact with other rocks. RAB drilling at extreme depth, if it encounters water, may rapidly clog the outside of the hole with debris, precluding removal of drill cuttings from the hole. This can be counteracted, however, with the use of "stabilizers" also known as "reamers", which are large cylindrical pieces of steel attached to the drill string, and made to perfectly fit the size of the hole being drilled. These have sets of rollers on the side, usually with tungsten buttons, that constantly break down cuttings being pushed upwards.

The use of high-powered air compressors, which push 900-1150 cfm of air at 300-350 psi down the hole also ensures drilling of a deeper hole up to ~1250 m due to higher air pressure which pushes all rock cuttings and any water to the surface. This, of course, is all dependent on the density and weight of the rock being drilled, and on how worn the drill bit is.

Air Core Drilling

Air core drilling and related methods use hardened steel or tungsten blades to bore a hole into unconsolidated ground. The drill bit has three blades arranged around the bit head, which cut the unconsolidated ground. The rods are hollow and contain an inner tube which sits inside the hollow outer rod barrel. The drill cuttings are removed by injection of compressed air into the hole via the annular area between the innertube and the drill rod. The cuttings are then blown back to surface up the inner tube where they pass through the sample separating system and are collected if needed. Drilling continues with the addition of rods to the top of the drill string. Air core drilling can occasionally produce small chunks of cored rock.

This method of drilling is used to drill the weathered regolith, as the drill rig and steel or tungsten blades cannot penetrate fresh rock. Where possible, air core drilling is preferred over RAB drilling as it provides a more representative sample. Air core drilling can achieve depths approaching 300 metres in good conditions. As the cuttings are removed inside the rods and are less prone to contamination compared to conventional drilling where the cuttings pass to the surface via outside return between the outside of the drill rod and the walls of the hole. This method is more costly and slower than RAB. This method of drilling was invented by Wallis Drilling a drilling company based in Perth, Western Australia.

Cable Tool Drilling

Cable tool rigs are a traditional way of drilling water wells. The majority of large diameter water supply wells, especially deep wells completed in bedrock aquifers, were completed using this drilling method. Although this drilling method has largely been supplanted in recent years by other, faster drilling techniques, it is still the most feasible drilling method for large diameter, deep bedrock wells, and in widespread use for small rural water supply wells. The impact of the drill bit fractures the rock and in many shale rock situations increases the water flow into a well over rotary.

Also known as ballistic well drilling and sometimes called "spudders", these rigs raise and drop a drill string with a heavy carbide tipped drilling bit that chisels through the rock by finely pulverizing the subsurface materials. The drill string is composed of the upper drill rods, a set of "jars" (inter-locking "sliders" that help transmit additional energy to the drill bit and assist in removing the bit if it is stuck) and the drill bit. During the drilling process, the drill string is periodically removed from the borehole and a bailer is lowered to collect the drill cuttings (rock fragments, soil, etc.). The bailer is a bucket-like tool with a trapdoor in the base. If the borehole is dry, water is added so that the drill cuttings will flow into the bailer. When lifted, the trapdoor closes and the cuttings are then raised and removed. Since the drill string must be raised and lowered to advance the boring, the casing (larger diameter outer piping) is typically used to hold back upper soil materials and stabilize the borehole.

Cable tool water well drilling rig in West Virginia.
These slow rigs have mostly been replaced by
rotary drilling rigs in the U.S.

Cable tool rigs are simpler and cheaper than similarly sized rotary rigs, although loud and very slow to operate. The world record cable tool well was drilled in New York to a depth of almost 12,000 feet (3,700 m). The common Bucyrus-Erie 22 can drill down to about 1,100 feet (340 m). Since cable tool drilling does not use air to eject the drilling chips like a rotary, instead using a cable strung bailer, technically there is no limitation on depth.

Cable tool rigs now are nearly obsolete in the United States. They are mostly used in Africa or Third-World countries. Being slow, cable tool rig drilling means increased wages for drillers. In the United States drilling wages would average around US$200 per day per man, while in Africa it is only US$6 per day per man, so a slow drilling machine can still be used in undeveloped countries with depressed wages. A cable tool rig can drill 25 feet (7.6 m) to 60 feet (18 m) of hard rock a day. A newer rotary drillcat top head rig equipped with down-the-hole (DTH) hammer can drill 500 feet (150 m) or more per day, depending on size and formation hardness.

Reverse Circulation (RC) Drilling

RC drilling is similar to air core drilling, in that the drill cuttings are returned to surface inside the rods. The drilling mechanism is a pneumatic reciprocating piston known as a "hammer" driving a tungsten-steel drill bit. RC drilling utilises much larger rigs and machinery and depths of up to 500 metres are routinely achieved. RC drilling ideally produces dry rock chips, as large air compressors dry the rock out ahead of the advancing drill bit. RC drilling is slower and costlier but achieves better penetration than RAB or air core drilling; it is cheaper than diamond coring and is thus preferred for most mineral exploration work.

Track mounted Reverse Circulation rig (side view).

Reverse circulation is achieved by blowing air down the rods, the differential pressure creating air lift of the water and cuttings up the "inner tube", which is inside each rod. It reaches the "divertor" at the top of the hole, then moves through a sample hose which is attached to the top of the "cyclone". The drill cuttings travel around the inside of the cyclone until they fall through an opening at the bottom and are collected in a sample bag.

Reverse Circulation Drilling set-up on Vertical
Travel Leads at the Port of La Rochelle, France.

The most commonly used RC drill bits are 5-8 inches (13–20 cm) in diameter and have round tungsten 'buttons' that protrude from the bit, which are required to drill through shale and abrasive rock. As the buttons wear down, drilling becomes slower and the rod string can potentially become bogged in the hole. This is a problem as trying to recover the rods may take hours and in some cases weeks. The rods and drill bits themselves are very expensive, often resulting in great cost to drilling companies when equipment is lost down the bore hole. Most companies will regularly re-grind the buttons on their drill bits in order to prevent this, and to speed up progress. Usually, when something is lost (breaks off) in the hole, it is not the drill string, but rather from the bit, hammer, or stabilizer to the bottom of the drill string (bit). This is usually caused by operator error, over-stressed metal, or adverse drilling conditions causing downhole equipment to get stuck in a part of the hole.

Although RC drilling is air-powered, water is also used to reduce dust, keep the drill bit cool, and assist in pushing cutting back upwards, but also when "collaring" a new hole. A mud called "Liqui-Pol" is mixed with water and pumped into the rod string, down the hole. This helps to bring up the sample to the surface by making the sand stick together. Occasionally, "Super-Foam" (a.k.a. "Quik-Foam") is also used, to bring all the very fine cuttings to the surface, and to clean the hole. When the drill reaches hard rock, a "collar" is put down the hole around the rods, which is normally PVC piping. Occasionally the collar may be made from metal casing. Collaring a hole is needed to stop the walls from caving in and bogging the rod string at the top of the hole. Collars may be up to 60 metres deep, depending on the ground, although if drilling through hard rock a collar may not be necessary.

Reverse circulation rig setups usually consist of a support vehicle, an auxiliary vehicle, as well as the rig itself. The support vehicle, normally a truck, holds diesel and water tanks for resupplying the rig. It also holds other supplies needed for maintenance on the rig. The auxiliary is a vehicle, carrying an auxiliary engine and a booster engine. These engines are connected to the rig by high pressure air hoses. Although RC rigs have their own booster and compressor to generate air pressure, extra power is needed which usually isn't supplied by the rig due to lack of space for these large engines. Instead, the engines are mounted on the auxiliary vehicle. Compressors on an RC rig have an output of around 1000 cfm at 500 psi (500 L·s^{-1} at 3.4 MPa). Alternatively, stand-alone air compressors which have an output of 900-1150cfm at 300-350 psi each are used in sets of 2, 3, or 4, which are all routed to the rig through a multi-valve manifold.

Diamond Core Drilling

Multi-combination drilling rig (capable of both diamond and reverse circulation drilling). Rig is currently set up for diamond drilling.

Diamond core drilling (exploration diamond drilling) uses an annular diamond-impregnated drill bit attached to the end of hollow drill rods to cut a cylindrical core of solid rock. The diamonds used to make diamond core bits are a variety of sizes, fine to microfine industrial grade diamonds, and the ratio of diamonds to metal used in the matrix affects the performance of the bits cutting ability in different types of rock formations. The diamonds are set within a matrix of varying hardness, from brass to high-grade steel. Matrix hardness, diamond size and dosing can be varied according to the rock which must be cut. The bits made with hard steel with a low diamond count are ideal for softer highly fractured rock while others made of softer steels and high diamond ratio are good for coring in hard solid rock. Holes within the bit allow water to be delivered to the cutting face. This provides three essential functions — lubrication, cooling, and removal of drill cuttings from the hole.

Diamond drilling is much slower than reverse circulation (RC) drilling due to the hardness of the ground being drilled. Drilling of 1200 to 1800 metres is common and at these depths, ground is mainly hard rock. Techniques vary among drill operators and what the rig they are using is capable of, some diamond rigs need to drill slowly to lengthen the life of drill bits and rods, which are very expensive and time consuming to replace at extremely deep depths. As a diamond drill rig cores deeper and deeper the time consuming part of the process is not cutting 5 to 10 more feet of rock core but the retrieval of the core with the wire line & overshot tool. Core samples are retrieved via the use of a core tube, a hollow tube placed inside the rod string and pumped with water until it locks into the core barrel. As the core is drilled, the core barrel slides over the core as it is cut. An "overshot" attached to the end of the winch cable is lowered inside the rod string and locks on to the backend (aka head assembly), located on the top end of the core barrel. The winch is retracted, pulling the core tube to the surface. The core does not drop out of the inside of the core tube when lifted because either a split ring core lifter or basket retainer allow the core to move into, but not back out of the tube.

Diamond core drill bits.

Once the core tube is removed from the hole, the core sample is then removed from the core tube and catalogued. The Driller's assistant unscrews the backend off the core tube using tube wrenches, then each part of the tube is taken and the core is shaken out into core trays. The core is washed, measured and broken into smaller pieces using a hammer or sawn through to make it fit into the sample trays. Once catalogued, the core trays are retrieved by geologists who then analyse the core and determine if the drill site is a good location to expand future mining operations.

Diamond rigs can also be part of a multi-combination rig. Multi-combination rigs are a dual setup rig capable of operating in either a reverse circulation (RC) and diamond drilling role (though not at the same time). This is a common scenario where exploration drilling is being performed in a very isolated location. The rig is first set up to drill as an RC rig and once the desired metres are drilled, the rig is set up for diamond drilling. This way the deeper metres of the hole can be drilled without moving the rig and waiting for a diamond rig to set up on the pad.

Direct Push Rigs

Direct push technology includes several types of drilling rigs and drilling equipment which advances a drill string by pushing or hammering without rotating the drill string. While this does not meet the proper definition of drilling, it does achieve the same result — a borehole. Direct push rigs include both cone penetration testing (CPT) rigs and direct push sampling rigs such as a PowerProbe or Geoprobe. Direct push rigs typically are limited to drilling in unconsolidated soil materials and very soft rock.

CPT rigs advance specialized testing equipment (such as electronic cones), and soil samplers using large hydraulic rams. Most CPT rigs are heavily ballasted (20 metric tons is typical) as a counter force against the pushing force of the hydraulic rams which are often rated up to 20 kN. Alternatively, small, light CPT rigs and offshore CPT rigs will use anchors such as screwed-in ground anchors to create the reactive force. In ideal conditions, CPT rigs can achieve production rates of up to 250–300 meters per day.

Direct push drilling rigs use hydraulic cylinders and a hydraulic hammer in advancing a hollow core sampler to gather soil and groundwater samples. The speed and depth of penetration is largely dependent on the soil type, the size of the sampler, and the weight and power of the rig. Direct push techniques are generally limited to shallow soil sample recovery in unconsolidated soil materials. The advantage of direct push technology is that in the right soil type it can produce a large number of high quality samples quickly and cheaply, generally from 50 to 75 meters per day. Rather than hammering, direct push can also be combined with sonic (vibratory) methods to increase drill efficiency.

Hydraulic Rotary Drilling

Oil well drilling utilises tri-cone roller, carbide embedded, fixed-cutter diamond, or diamond-impregnated drill bits to wear away at the cutting face. This is preferred because there is no need to return intact samples to surface for assay as the objective is to reach a formation containing oil or natural gas. Sizable machinery is used, enabling depths of several kilometres to be penetrated. Rotating hollow drill pipes carry down bentonite and barite infused drilling muds to lubricate, cool, and clean the drilling bit, control downhole pressures, stabilize the wall of the borehole and remove drill cuttings. The mud travels back to the surface around the outside of the drill pipe, called the annulus. Examining rock chips extracted from the mud is known as mud logging. Another form of well logging is electronic and is commonly employed to evaluate the existence of possible oil and gas deposits in the borehole. This can take place while the well is being drilled, using Measurement While Drilling tools, or after drilling, by lowering measurement tools into the newly drilled hole.

The rotary system of drilling was in general use in Texas in the early 1900s. It is a modification of one invented by Fauvelle in 1845, and used in the early years of the oil industry in some of the oil-producing countries in Europe. Originally pressurized water was used instead of mud, and was almost useless in hard rock before the diamond cutting bit. The main breakthrough for rotary drilling came in 1901, when Anthony Francis Lucas combined the use of a steam-driven rig and of mud instead of water in the Spindletop discovery well.

The drilling and production of oil and gas can pose a safety risk and a hazard to the environment from the ignition of the entrained gas causing dangerous fires and also from the risk of oil leakage polluting water, land and groundwater. For these reasons, redundant safety systems and highly trained personnel are required by law in all countries with significant production.

Automated Drill Rig

An automated drill rig (ADR) is an automated full-sized walking land-based drill rig that drills long lateral sections in horizontal wells for the oil and gas industry. ADRs are agile rigs that can move from pad to pad to new well sites faster than other full-sized drilling rigs. Each rig costs about $25 million. ADR is used extensively in the Athabasca oil sands. According to the "Oil Patch Daily News", "Each rig will generate 50,000 man-hours of work during the construction phase and upon completion, each operating rig will directly and indirectly employ more than 100 workers." Compared to conventional drilling rigs", Ensign, an international oilfield services contractor based in Calgary, Alberta, that makes ADRs claims that they are "safer to operate, have "enhanced controls intelligence," "reduced environmental footprint, quick mobility and advanced communications between field and office." In June 2005 the first specifically designed slant automated drilling rig (ADR), Ensign Rig No. 118, for steam assisted gravity drainage (SAGD) applications was mobilized by Deer Creek Energy Limited, a Calgary-based oilsands company.

Limits of the Technology

Drill technology has advanced steadily since the 19th century. However, there are several basic limiting factors which will determine the depth to which a bore hole can be sunk.

All holes must maintain outer diameter; the diameter of the hole must remain wider than the diameter of the rods or the rods cannot turn in the hole and progress cannot continue. Friction caused by the drilling operation will tend to reduce the outside diameter of the drill bit. This applies to all drilling methods, except that in diamond core drilling the use of thinner rods and casing may permit the hole to continue. Casing is simply a hollow sheath which protects the hole against collapse during drilling, and is made of metal or PVC. Often diamond holes will start off at a large diameter and when outside diameter is lost, thinner rods put down inside casing to continue, until finally the hole becomes too narrow. Alternatively, the hole can be reamed; this is the usual practice in oil well drilling where the hole size is maintained down to the next casing point.

For percussion techniques, the main limitation is air pressure. Air must be delivered to the piston at sufficient pressure to activate the reciprocating action, and in turn drive the head into the rock with sufficient strength to fracture and pulverise it. With depth, volume is added to the in-rod string, requiring larger compressors to achieve operational pressures. Secondly, groundwater is ubiquitous, and increases in pressure with depth in the ground. The air inside the rod string must be pressurised enough to overcome this water pressure at the bit face. Then, the air must be able to carry the rock fragments to surface. This is why depths in excess of 500 m for reverse circulation drilling are rarely achieved, because the cost is prohibitive and approaches the threshold at which diamond core drilling is more economic.

Diamond drilling can routinely achieve depths in excess of 1200 m. In cases where money is no issue, extreme depths have been achieved, because there is no requirement to overcome water pressure. However, water circulation must be maintained to return the drill cuttings to surface, and more importantly to maintain cooling and lubrication of the cutting surface of the bit; while at the same time reduce friction on the steel walls of the rods turning against the rock walls of the hole. When water return is lost the rods will vibrate, this is called "rod chatter", and that will damage the drill rods, and crack the joints.

Without sufficient lubrication and cooling, the matrix of the drill bit will soften. While diamond is the hardest substance known, at 10 on the Mohs hardness scale, it must remain firmly in the matrix to achieve cutting. Weight on bit, the force exerted on the cutting face of the bit by the drill rods in the hole above the bit, must also be monitored.

Causes of Deviation

Most drill holes deviate slightly from their planned trajectory. This is because of the torque of the turning bit working against the cutting face, because of the flexibility of the steel rods and especially the screw joints, because of reaction to foliation and structure within the rock, and because of refraction as the bit moves into different rock layers of varying resistance. Additionally, inclined holes will tend to deviate upwards because the drill rods will lie against the bottom of the bore, causing the drill bit to be slightly inclined from true. It is because of deviation that drill holes must be surveyed if deviation will impact the usefulness of the information returned. Sometimes the surface location can be offset laterally to take advantage of the expected deviation tendency, so the bottom of the hole will end up near the desired location. Oil well drilling commonly uses a process of controlled deviation called directional drilling (e.g., when several wells are drilled from one surface location).

Rig Equipment

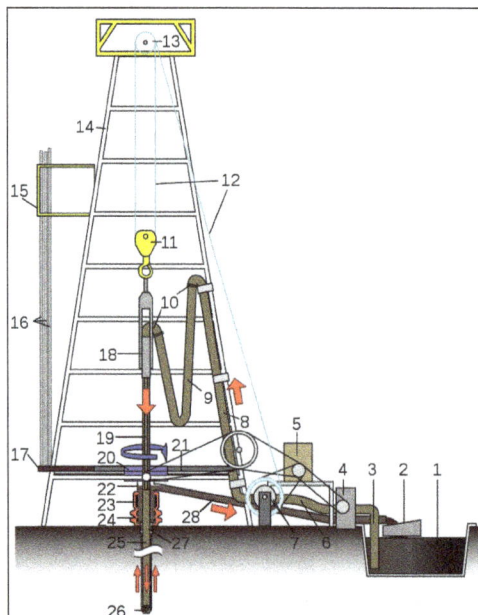
Simple diagram of a drilling rig and its basic operation.

Drilling rigs typically include at least some of the following items:

Blowout Preventers: (BOPs)

The equipment associated with a rig is to some extent dependent on the type of rig but (#23 & #24) are devices installed at the wellhead to prevent fluids and gases from unintentionally escaping from the borehole. #23 is the annular (often referred to as the "Hydril", which is one manufacturer) and #24 is the pipe rams and blind rams. In the place of #24 Variable bore rams or VBRs can be used. These offer the same pressure and sealing capacity found in standard pipe rams, while offering the versatility of sealing on various sizes of drill pipe, production tubing and casing without changing standard pipe rams. Normally VBRs are used when utilizing a tapered drill string (when different size drill pipe is used in the complete drill string):

- Centrifuge: An industrial version of the device that separates fine silt and sand from the drilling fluid.

- Solids control: Solids control equipment is for preparing drilling mud for the drilling rig.

- Chain tongs: Wrench with a section of chain, that wraps around whatever is being tightened or loosened. Similar to a pipe wrench.

- Degasser: A device that separates air and/or gas from the drilling fluid.

- Desander / desilter: Contains a set of hydrocyclones that separate sand and silt from the drilling fluid.

- Drawworks: (#7) is the mechanical section that contains the spool, whose main function is to reel in/out the drill line to raise/lower the traveling block (#11).

- Drill bit: (#26) is a device attached to the end of the drill string that breaks apart the rock being drilled. It contains jets through which the drilling fluid exits.

- Drill pipe: (#16) joints of hollow tubing used to connect the surface equipment to the bottom hole assembly (BHA) and acts as a conduit for the drilling fluid. In the diagram, these are "stands" of drill pipe which are 2 or 3 joints of drill pipe connected together and "stood" in the derrick vertically, usually to save time while tripping pipe.

- Elevators: A gripping device that is used to latch to the drill pipe or casing to facilitate the lowering or lifting (of pipe or casing) into or out of the borehole.

- Mud motor: A hydraulically powered device positioned just above the drill bit used to spin the bit independently from the rest of the drill string.

- Mud pump: (#4) reciprocal type of pump used to circulate drilling fluid through the system.

- Mud tanks: (#1) often called mud pits, provides a reserve store of drilling fluid until it is required down the wellbore.

- Rotary table: (#20) rotates the drill string along with the attached tools and bit.

- Shale shaker: (#2) separates drill cuttings from the drilling fluid before it is pumped back down the borehole.

Occupational Safety

Drilling rigs create some safety challenges for those who work on them. One safety concern is the use of seatbelts for workers driving between two locations. Motor vehicle fatalities on the job for these workers is 8.5 times the rate of the rest of the US working population, which can be attributed to the low rate of seatbelt use.

Drill String

A drill string on a drilling rig is a column, or string, of drill pipe that transmits drilling fluid (via the mud pumps) and torque (via the kelly drive or top drive) to the drill bit. The term is loosely applied to the assembled collection of the smuggler pool, drill collars, tools and drill bit. The drill string is hollow so that drilling fluid can be pumped down through it and circulated back up the annulus (the void between the drill string and the casing/open hole).

Drill String Components

The drill string is typically made up of three sections:

- Bottom hole assembly (BHA).
- Transition pipe, which is often heavyweight drill pipe (HWDP).
- Drill pipe.

Bottom Hole Assembly (BHA)

The Bottom Hole Assembly (BHA) is made up of: a drill bit, which is used to break up the rock formations; drill collars, which are heavy, thick-walled tubes used to apply weight to the drill bit; and drilling stabilizers, which keep the assembly centered in the hole. The BHA may also contain other components such as a downhole motor and rotary steerable system(RSS), measurement while drilling (MWD), and logging while drilling (LWD) tools. The components are joined together using rugged threaded connections. Short "subs" are used to connect items with dissimilar threads.

Transition Pipe

Heavyweight drill pipe (HWDP) may be used to make the transition between the drill collars and drill pipe. The function of the HWDP is to provide a flexible transition between the drill collars and the drill pipe. This helps to reduce the number of fatigue failures seen directly above the BHA. A secondary use of HWDP is to add additional weight to the drill bit. HWDP is most often used as weight on bit in deviated wells. The HWDP may be directly above the collars in the angled section of the well, or the HWDP may be found before the kick off point in a shallower section of the well.

Drill Pipe

Drill pipe makes up the majority of the drill string back up to the surface. Each drill pipe comprises a long tubular section with a specified outside diameter (e.g. 3 1/2 inch, 4 inch, 5 inch, 5 1/2 inch, 5 7/8 inch, 6 5/8 inch). At each end of the drill pipe tubular, larger-diameter portions called the tool joints are located. One end of the drill pipe has a male ("pin") connection whilst the other has a female ("box") connection. The tool joint connections are threaded which allows for the mating of each drill pipe segment to the next segment.

Running a Drill String

Most components in a drill string are manufactured in 31 foot lengths (range 2) although they can also be manufactured in 46 foot lengths (range 3). Each 31 foot component is referred to as a joint. Typically 2, 3 or 4 joints are joined together to make a stand. Modern onshore rigs are capable of handling ~90 ft stands (often referred to as a triple).

Pulling the drill string out of or running the drill string into the hole is referred to as tripping. Drill pipe, HWDP and collars are typically racked back in stands in to the monkeyboard which is a component of the derrick if they are to be run back into the hole again after, say, changing the bit. The disconnect point ("break") is varied each subsequent round trip so that after three trips every connection has been broken apart and later made up again with fresh pipe dope applied.

Stuck Drill String

A stuck drill string can be caused by many situations:

- Packing-off due to cuttings settling back into the wellbore when circulation is stopped.

- Differentially when there is a large difference between formation pressure and wellbore pressure. The drill string is pushed against one side of the well bore. The force required to pull the string along the wellbore in this occurrence is a function of the total contact surface area, the pressure difference and the friction factor.

- Keyhole sticking occurs mechanically as a result of pulling up into doglegs when tripping.

- Adhesion due to not moving it for a significant amount of time.

Once the tubular member is stuck, there are many techniques used to extract the pipe. The tools and expertise are normally supplied by an oilfield service company. Two popular tools and techniques are the oilfield jar and the surface resonant vibrator. Below is a history of these tools along with how they operate.

Jars

The mechanical success of cable tool drilling has greatly depended on a device called jars, invented by a spring pole driller, William Morris, in the salt well days of the 1830s. Little is known about Morris except for his invention and that he listed Kanawha County (now in West Virginia) as his address. Morris received US 2243 for this unique tool in 1841 for artesian well drilling. Later, using jars, the cable tool system was able to efficiently meet the demands of drilling wells for oil.

The jars were improved over time, especially at the hands of the oil drillers, and reached the most useful and workable design by the 1870s, due to another US 78958 received in 1868 by Edward Guillod of Titusville, Pennsylvania, which addressed the use of steel on the jars' surfaces that were subject to the greatest wear. Many years later, in the 1930s, very strong steel alloy jars were made.

8 inch drilling jar (red and white) on casings.

A set of jars consisted of two interlocking links which could telescope. In 1880 they had a play of about 13 inches such that the upper link could be lifted 13 inches before the lower link was engaged. This engagement occurred when the cross-heads came together.Today, there are two primary types, hydraulic and mechanical jars. While their respective designs are quite different, their operation is similar. Energy is stored in the drillstring and suddenly released by the jar when it fires. Jars can be designed to strike up, down, or both. In the case of jarring up above a stuck bottomhole assembly, the driller slowly pulls up on the drillstring but the BHA does not move. Since the top of the drillstring is moving up, this means that the drillstring itself is stretching and storing energy. When the jars reach their firing point, they suddenly allow one section of the jar to move axially relative to a second, being pulled up rapidly in much the same way that one end of a stretched spring moves when released. After a few inches of movement, this moving section slams into a steel shoulder, imparting an impact load.

In addition to the mechanical and hydraulic versions, jars are classified as drilling jars or fishing jars. The operation of the two types is similar, and both deliver approximately the same impact blow, but the drilling jar is built such that it can better withstand the rotary and vibrational loading associated with drilling. Jars are designed to be reset by simple string manipulation and are capable of repeated operation or firing before being recovered from the well. Jarring effectiveness is determined by how rapidly you can impact weight into the jars. When jarring without a compounder or accelerator you rely only on pipe stretch to lift the drill collars upwards after the jar releases to create the upwards impact in the jar. This accelerated upward movement will often be reduced by the friction of the working string along the sides of the well bore, reducing the speed of upwards movement of the drill collars which impact into the jar. At shallow depths jar impact is not achieved because of lack of pipe stretch in the working string.

When pipe stretch alone cannot provide enough energy to free a fish, compounders or accelerators are used. Compounders or accelerators are energized when you over pull on the working string and compress a compressible fluid through a few feet of stroke distance and at the same time activate the fishing jar. When the fishing jar releases the stored energy in the compounder/acclerator lifts the drill collars upwards at a high rate of speed creating a high impact in the jar.

System Dynamics of Jars

Jars rely on the principle of stretching a pipe to build elastic potential energy such that when the jar trips it relies on the masses of the drill pipe and collars to gain velocity and subsequently strike the anvil section of jar. This impact results in a force, or blow, which is converted into energy.

Surface Resonant Vibrators

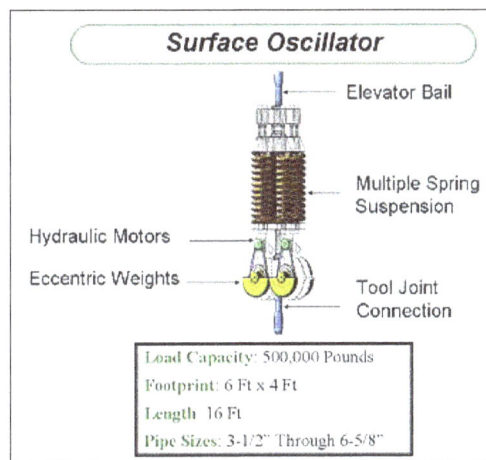

Oilfield Surface Resonant Vibrator.

The concept of using vibration to free stuck objects from a wellbore originated in the 1940s, and probably stemmed from the 1930s use of vibration to drive piling in the Soviet Union. The early use of vibration for driving and extracting piles was confined to low frequency operation; that is, frequencies less than the fundamental resonant frequency of the system and consequently, although effective, the process was only an improvement on conventional hammer equipment. Early patents and teaching attempted to explain the process and mechanism involved, but lacked a certain degree of sophistication. In 1961, A. G. Bodine obtained US 2972380 that was to become the "mother patent" for oil field tubular extraction using sonic techniques. Mr. Bodine introduced the concept of resonant vibration that effectively eliminated the reactance portion of mechanical impedance, thus leading to the means of efficient sonic power transmission. Subsequently, Mr. Bodine obtained additional patents directed to more focused applications of the technology.

The first published work on this technique was outlined in a 1987 Society of Petroleum Engineers (SPE) paper presented at the International Association of Drilling Contractors in Dallas, Texas detailing the nature of the work and the operational results that were achieved. The cited work involving liner, tubing, and drill pipe extraction and was very successful. Reference Two presented at the Society of Petroleum Engineers Annual Technical Conference and Exhibition in Anaheim, California, November, 2007 explains the resonant vibration theory in more detail as well as its use in extracting long lengths of mud stuck tubulars.

System Dynamics of Surface Resonant Vibrators

Surface Resonant Vibrators rely on the principle of counter rotating eccentric weights to impart a sinusoidal harmonic motion from the surface into the work string at the surface. The frequency of rotation, and hence vibration of the pipe string, is tuned to the resonant frequency of the system. The system is defined as the surface resonant vibrator, pipe string, fish and retaining media. The resultant forces imparted to the fish is based on the following logic:

- The delivery forces from the surface are a result of the static overpull force from the rig, plus the dynamic force component of the rotating eccentric weights.

- Depending on the static overpull force component, the resultant force at the fish can be either tension or compression due to the sinusoidal force wave component from the oscillator.

- Initially during startup of a vibrator, some force is necessary to lift and lower the entire load mass of the system. When the vibrator tunes to the resonant frequency of the system, the reactive load impedance cancels out to zero by virtue of the inductance reactance (mass of the system) equaling the compliance or stiffness reactance (elasticity of the tubular). The remaining impedance of the system, known as the resistive load impedance, is what is retaining the stuck pipe.

- During resonant vibration, a longitudinal sine wave travels down the pipe to the fish with an attendant pipe mass that is equal to a quarter wavelength of the resonant vibrating frequency.

- A phenomenon known as fluidization of soil grains takes place during resonant vibration whereby the granular material constraining the stuck pipe is transformed into a fluidic state that offers little resistance to movement of bodies through the media. In effect, it takes on the characteristics and properties of a liquid.

- During pipe vibration, Dilation and Contraction of the pipe body, known as Poisson's ratio, takes place such that when the stuck pipe is subjected to axial strain due to stretching, its diameter will contract. Similarly, when the length of pipe is compressed, its diameter will expand. Since a length of pipe undergoing vibration experiences alternate tensile and compressive forces as waves along its longitudinal axis (and therefore longitudinal strains), its diameter will expand and contract in unison with the applied tensile and compressive waves. This means that for alternate moments during a vibration cycle the pipe may actually be physically free of its bond.

Drill Cuttings

Drill cuttings are the broken bits of solid material removed from a borehole drilled by rotary, percussion, or auger methods and brought to the surface in the drilling mud. Boreholes drilled in this way include oil or gas wells, water wells, and holes drilled for geotechnical investigations or mineral exploration.

The drill cuttings are commonly examined to make a record (a well log) of the subsurface materials penetrated at various depths. In the oil industry, this is often called a mud log.

Drill cuttings are produced as the rock is broken by the drill bit advancing through the rock or soil; the cuttings are usually carried to the surface by drilling fluid circulating up from the drill bit. Drill cuttings can be separated from liquid drilling fluid by shale shakers, by centrifuges, or by cyclone separators, the latter also being effective for air drilling. In cable-tool drilling, the drill cuttings are periodically bailed out of the bottom of the hole. In auger drilling, cuttings are carried to the surface on the auger flights.

One drilling method that does not produce drill cuttings is core drilling, which instead produces solid cylinders of rock or soil.

Management of Drill Cuttings

Drill cuttings carried by mud (drilling fluid) are usually retrieved at the surface of the platform where they go through shakers or vibrating machines to separate the cuttings from the drilling fluid, this process allows the circulating fluid to renter the drilling process.

Samples from the cuttings are then studied by mud loggers and wellsite geologist. In the Oil and Gas industry the operator will likely require a set of samples for further analysis in their labs. Many national regulations stipulate that for any well drilled a set of samples must be archived with a national body. For example in the case of the UK with the British Geological Survey (BGS).

The bulk of the cuttings require disposal. The methodology is dependent on the type of drilling fluid use. For water based drilling fluid (WBM) with no particular dangerous additives, the cuttings can be dumped overboard (in offshore senario). If however an oil based drilling fluid (OBM) is used then the cuttings must be processed before disposal. Either in skips and transported to a dedicated facility (aka skip and ship), or now there are mobile plants that can process them at the rigsite effectively burning off the drilling fluid contamination. This saves the logistics and cost of transporting such quantities of cuttings. Although possibly thought of as a uninteresting topic, if in a skip and ship scenario, the dependency on crane operations to move skips can lead to situations whereby bad weather halts drilling as the cuttings handling cannot continue.

Disposal as Waste

Burial

Burial is the placement of waste in man-made or natural excavations, such as pits or landfills. Burial is the most common onshore disposal technique used for disposing of drilling wastes (mud and cuttings). Generally, the solids are buried in the same pit (the reserve pit) used for collection and temporary storage of the waste mud and cuttings after the liquid is allowed to evaporate. Pit burial is a low-cost, low-tech method that does not require wastes to be transported away from the well site, and, therefore, is very attractive to many operators.

Burial may be the most misunderstood or misapplied disposal technique. Simply pushing the walls of the reserve pit over the drilled cuttings is generally not acceptable. The depth or placement of the burial cell is important. A moisture content limit should be established on the buried cuttings, and the chemical composition should be determined. Onsite pit burial may not be a good choice for wastes that contain high concentrations of oil, salt, biologically available metals, industrial chemicals, and other materials with harmful components that could migrate from the pit and contaminate usable water resources.

In some oil field areas, large landfills are operated to dispose of oil field wastes from multiple wells. Burial usually results in anaerobic conditions, which limits any further degradation when compared with wastes that are land-farmed or land-spread, where aerobic conditions predominate.

Application to Land Surfaces

The objective of applying drilling wastes to the land is to allow the soil's naturally occurring microbial population to metabolize, transform, and assimilate waste constituents in place. Land application is a form of bioremediation are described in a separate fact sheet.

Several terms are used to describe this waste management approach, which can be considered both treatment and disposal. In general, *land farming* refers to the repeated application of wastes to the soil surface, whereas *land spreading* and *land treatment* are often used interchangeably to describe the one-time application of wastes to the soil surface. Some practitioners do not follow the same terminology convention, and may interchange all three terms. Readers should focus on the technologies rather than on the specific names given to each process.

Optimal land application techniques balance the additions of waste against a soil's capacity to assimilate the waste constituents without destroying soil integrity, creating subsurface soil contamination problems, or causing other adverse environmental impacts.

Land Farming

The exploration and production industry has used land farming to treat oily petroleum industry wastes for years. Land farming is the controlled and repeated application of wastes to the soil surface, using microorganisms in the soil to naturally biodegrade hydrocarbon constituents, dilute and attenuate metals, and transform and assimilate waste constituents.

Land farming can be a relatively low-cost drilling waste management approach. Some studies indicate that land farming does not adversely affect soils and may even benefit certain sandy soils by increasing their water-retaining capacity and reducing fertilizer losses. Inorganic compounds and metals are diluted in the soil, and may also be incorporated into the matrix (through chelation, exchange reactions, covalent bonding, or other processes) or may become less soluble through oxidation, precipitation, and pH effects. The attenuation of heavy metals (or the taking up of metals by plants) can depend on clay content and cation-exchange capacity.

Optimizing Land Farm Operations: The addition of water, nutrients, and other amendments (e.g., manure, straw) can increase the biological activity and aeration of the soil, thereby preventing the development of conditions that might promote leaching and mobilization of inorganic contaminants. During periods of extended dry conditions, moisture control may also be needed to minimize dust.

Periodic tillage of the mixture (to increase aeration) and nutrient additions to the waste-soil mixture can enhance aerobic biodegradation of hydrocarbons. After applying the wastes, hydrocarbon concentrations are monitored to measure progress and determine the need for enhancing the biodegradation processes. Application rates should be controlled to minimize the potential for runoff.

Pretreating the wastes by composting and activating aerobic biodegradation by regular turning (windrows) or by forced ventilation (biopiles) can reduce the amount of acreage required for land farming.

Drilling Waste Land Farm Example: In 1995, HS Resources, an oil and gas company operating in Colorado, obtained a permit for a noncommercial land farm to treat and recycle the company's nonhazardous oil field wastes, including drilling muds. At the land farm, wastes mixed with soil contaminated with hydrocarbons from other facilities are spread in a layer one foot thick or less. Natural bacterial action is enhanced through occasional addition of commercial fertilizers, monthly tilling (to add oxygen), and watering (to maintain 10–15% moisture content). Treatment is considered complete when hydrocarbon levels reach concentrations specified by regulatory agencies; not all agencies employ the same acceptability standards. Water and soil are monitored periodically to confirm that no adverse soil or groundwater impacts have occurred, and records of the source and disposition of the remediated soil are maintained. Estimated treatment costs, which include transportation, spreading, amendments, and monitoring, are about $4–5 per cubic yard. When the treated material is recycled as backfill, net costs are about $1 per cubic yard. Capital costs (not included in the treatment cost estimates) were recovered within the first eight months of operation.

Implementation Considerations: Advantages of land farming include its simplicity and low capital cost, the ability to apply multiple waste loadings to the same parcel of land, and the potential to improve soil conditions. Concerns associated with land farming are its high maintenance costs (e.g., for periodic land tilling, fertilizer); potentially large land requirements; and required analysis, testing, demonstration, and monitoring. Elevated concentrations of hydrocarbons in drilling wastes can limit the application rate of a waste on a site.

Wastes containing salt must also be applied to soil only with care. Salt, unlike hydrocarbons, cannot biodegrade but may accumulate in soils, which have a limited capacity to accept salts. If salt levels become too high, the soils may be damaged and treatment of hydrocarbons can be inhibited. Salts are soluble in water and can be managed. Salt management is part of prudent operation of a land farm.

Another concern with land farming is that while lower molecularweight petroleum compounds biodegrade efficiently, higher molecular weight compounds biodegrade more slowly. This means that repeated applications can lead to accumulation of high molecular weight compounds. At high concentrations, these recalcitrant constituents can increase soil-water repellency, affect plant growth, reduce the ability of the soil to support a diverse community of organisms, and render the land farm no longer usable without treatment or amendment. Recent studies have supported the idea that field-scale additions of earthworms with selected organic amendments may hasten the long-term recovery of conventionally treated petroleum contaminated soil. The burrowing and feeding activities of earthworms create space and allow food resources to become available to other soil organisms that would be unable to survive otherwise.

When considering land farming as a waste management option, several items should be considered. These include site topography, site hydrology, neighboring land use, and the physical (texture and bulk density) and chemical composition of the waste and the resulting waste-soil mixture. Wastes that contain large amounts of oil and various additives may have diverse effects on parts of the food chain. Constituents of particular concern include pH, nitrogen (total mass), major soluble ions (Ca, Mg, Na, Cl), electrical conductivity, total metals, extractable organic halogens, oil content, and hydrocarbons. Oil-based muds typically utilize an emulsified phase of 20 to 35 percent by weight $CaCl_2$ brine. This salt can be a problem in some areas, such as some parts of Canada, the mid-continent, and the Rocky Mountains. For this reason, alternative mud systems have emerged that use an environmentally preferred beneficial salt, such as calcium nitrate or potassium sulfate, as the emulsified internal water phase.

Wastes that contain significant levels of biologically available heavy metals and persistent toxic compounds are not good candidates for land farming, as these substances can accumulate in the soil to a level that renders the land unfit for further use (E&P Forum 1993). (Site monitoring can help ensure such accumulation does not occur.) Land farms may require permits or other approvals from regulatory agencies, and, depending on soil conditions, some land farms may require liners and/or groundwater monitoring wells.

Land Treatment

In land treatment (also known as land spreading), the processes are similar to those in land farming, where natural soil processes are used to biodegrade the organic constituents in the waste. However, in land treatment, a one-time application of the waste is made to a parcel of land. The objective is to dispose of the waste in a manner that preserves the subsoil's chemical, biological, and physical properties by limiting the accumulation of contaminants and protecting the quality of surface and groundwater. The land spreading area is determined on the basis of a calculated loading rate that considers the absolute salt concentration, hydrocarbon concentration, metals concentration, and pH level after mixing with the soil. The drilling waste is spread on the land and incorporated into the upper soil zone (typically upper 6–8 inches of soil) to enhance hydrocarbon volatization and biodegradation. The land is managed so that the soil system can degrade, transport, and assimilate the waste constituents. Each land treatment site is generally used only once.

Optimizing Land Treatment Operations: Addition of water, nutrients, and other amendments (e.g., manure, straw) can increase the biological activity and aeration of the soil and prevent the development of conditions that might promote leaching and mobilization of inorganic contaminants. During periods of extended dry conditions, moisture control may also be needed to minimize dust. Periodic tillage of the mixture (to increase aeration) and nutrient additions to the waste soil mixture can enhance aerobic biodegradation of hydrocarbons, although in practice not all land treatment projects include repeated tilling. After applying the wastes, hydrocarbon concentrations may be monitored to measure progress and determine the need for enhancing the biodegradation processes.

Implementation Considerations: Because land spreading sites receive only a single application of waste, the potential for accumulation of waste components in the soil is reduced (as compared with land farming, where waste is applied repeatedly). Although liners and monitoring of leachate are typically not required at land treatment sites, site topography, hydrology, and the physical and chemical composition of the waste and resultant waste-soil mixture should be assessed, with waste application rates controlled to minimize the possibility of runoff.

Experiments conducted in France showed that after spreading oil-based mud cuttings on farmland, followed by plowing, tilling, and fertilizing, approximately 10% of the initial quantity of the oil remained in the soil. Phytotoxic effects on seed germination and sprouting were not observed, but corn and wheat crop yields decreased by 10%. Yields of other crops were not affected. The percentage of hydrocarbon reduction and crop yield performance will vary from site to site depending on many factors (e.g., length of time after application, type of hydrocarbon, soil chemistry, temperature).

Land spreading costs are typically $2.50 to $3.00 per barrel of water-based drilling fluids not contaminated with oil, and they could be higher for oily wastes containing salts. Costs also depend on sampling and analytical requirements.

Advantages of land spreading are the low treatment cost and the possibility that the approach could improve soil characteristics. Land spreading is most effectively used for drilling wastes that have low levels of hydrocarbons and salts. Potential concerns include the need for large land areas; the relatively slow degradation process (the rate of biodegradation is controlled by the inherent biodegradation properties of the waste constituents, soil temperature, soil-water content, and contact between the microorganisms and the wastes); and the need for analyses, tests, and demonstrations. Also, high concentrations of soluble salts or metals can limit the use of land spreading.

When evaluating land spreading as a drilling waste management option, several items should be considered. These include area-wide topographical and geological features; current and likely future activities around the disposal site; hydrogeologic data (location, size, and direction of flow for existing surface water bodies and fresh or usable aquifers); natural or existing drainage patterns; nearby environmentally sensitive features such as wetlands, urban areas, historical or archeological sites, and protected habitats; the presence of endangered species; and potential air quality impacts. In addition, historical rainfall distribution data should be reviewed to establish moisture requirements for land spreading and predict net evaporation rates. Devices needed to control water flow into, onto, or from facility systems should be identified. Wastes should be characterized during the evaluation; drilling wastes with high levels of hydrocarbons and salts may not be appropriate for land spreading.

Recycling

Some cuttings are being beneficially reused. Before the cuttings can be reused or recycled, it may be necessary to follow steps to ensure the hydrocarbon and chloride content is lowered within the governing bodies standards for reuse.

Reuse of cuttings through road spreading is permitted in some areas. However it depends on permission from not only the governing agency, but also the land owner.

Drill cutting can also be recycled for use as a bulk particulate solid construction materials such as road base for site roads and pads. The cuttings must first be screened and dried, before being processes in a pugmill or similar type mixing method.

Drilling wastes can also be recycled for use as a major constituent of mixes for making substantially monolithic specialized civil engineering concrete structures of large size, such as roads and drilling pads.

Types of Well Logging

There are many different methods of well logging that can be employed for solving various problems connected with geophysical exploration. These methods are based on different physical principles and have different operational techniques.

Electrical Logging

The most common techniques of geophysical well logging utilize variations in electrical properties in geological formations and the drilling fluid. Logical subdivisions of electrical logging include spontaneous potential (SP), single point resistance (SPR), resistivity (Normal, Lateral) and micro normal.

Spontaneous Potential Logging

The spontaneous potential log (SP) is a record of potentials or voltages that develop at the contacts between shale or clay beds and a sand aquifer, where they are penetrated by a borehole. The natural flow of current and the spontaneous potential curve that would be produced under the salinity conditions. The SP measuring equipment consists of a lead electrode in the well connected through a milli voltmeter to a second lead electrode that is grounded at the land surface.

The spontaneous potential is a function of the chemical activates of fluids in the borehole and adjacent rocks, the temperature, and the type and quantity of clay present. The chief source of SP in a borehole is electrochemical and electro kinetic or streaming potentials. Electrochemical effects probably are the most significant contributor; they can subdivide into membrane and liquid junction potentials. Both these effects are the result of the migration of ions from concentrated to more dilute solution, and they are mostly affected by clay, that decreases negative (anion) mobility. Membrane potentials are developed from formation water to adjacent shale, for fluid in the borehole; a three component system. Liquid junction potentials are those developed between the filtrate in the invaded zone and the formation water. If the fluid column in the borehole is more saline than the water in the aquifer, current flow and the log will be reversed. Streaming potentials are caused by the movement of an electrolyte through permeable media. They are substantial at depth intervals where water is moving in or out of the hole.

The SP logs have been used widely in determining lithology, bed thickness and the salinity of the formation water. Lithologic contacts are located on spontaneous potential logs at the point of curve inflection, where the current density is maximum. When the response is typical, a line can be drawn through the positive spontaneous potential values recorded in shale base line and a parallel line may be drawn through negative values that represent intervals of the sand base line containing little clay. If salinity and the composition of the borehole and the interstitial fluids are constant through the logged interval, shale and sand lines will vertically on the log; however, this is not common in water wells. Where the individual beds are thick enough these lines can be used to calculate sand-shale ratio or to calculate net thickness of each. The shale fraction is proportional to the relative spontaneous potential deflection between sand and shale beds. In sodium chloride type saline water, the following relation is used to calculate resistivity of formation water, R_w from log.

$$SP = -K' \, Log \, (R_M/R_w)$$

Where, SP= log deflection, in mill volts;

K'= 60+0.133T;

T= Borehole temperature in $^\circ$F;

R_m = Resistivity of borehole fluid in ohm-m;

and R_W = Resistivity of formation water in ohm-m.

The SP deflection is read from a log at the thick sand bed; R_m is measured by a mud cell. Temperature may obtain from a SP log. This equation cannot hold good always when the ground waters are dominated by chemical constituents other than sodium chloride.

Single Point Resistance Logging (SPR)

Single point resistance (SPR) logging has been one of the most widely used in Coal exploration, for identification of lithological information.

Ohm's law provides the basic principle for all logging devices that measure resistance, resistivity or conductivity. The resistance of any medium depends not only on its composition but also on the cross sectional area and length of the path through that medium. Single point resistance systems measure the resistance, in ohms, between an electrode in the well and an electrode at the land surface.

A Constant AC is supplied by a generator so that resistance is directly proportional to the potential read in milli volts. In a volt meter the unknown resistance is connected in series with a meter and a battery. The volume of investigation of the SPR probe is small about 5 to 10 times the electrode diameter.

The single point resistance log is useful for obtaining information about lithology. They have a significant advantage over multi electrode logs, because they do not have reversals as a result of bed thickness effects. SPR logs deflect in the proper direction in response to the resistivity of materials adjacent to the electrode, regardless of the bed thickness; thus they have a variable vertical resolution.

Resistivity Logging

The principles of electrical resistivity logging are similar to the surface electrical profiling method. Generally a four electrode system AMNB is used, where A and B are the current electrodes through which artificial energizing filed is sent and the potential difference between M and N (potential electrodes) are measured. In logging, commonly three of them are utilized as moving or in borehole electrodes. The assemblage of these three electrodes is called the measuring/sensing device termed as the sonde. The fourth electrode, either for potential or current is planted on the ground or placed in a mud pit on the surface close to the opening of well. A generator or battery is used to supply the current to the electrodes .The recording instruments with cable winch are installed in a logging truck. The measurements are taken while the sonde is drawn up from the borehole.

Types of Conventional Sondes

In conventional electrical resistivity logging, three electrodes are set up on the measuring device (sonde). The placement of current and potential electrodes will decide the type of sonde.

When one current and one potential electrode is kept close to each other compared to the distance from another current/potential on the sonde, it is called as the normal or potential sonde. It is sometimes termed as non-paired electrode set up.

If a pair of current or potential electrodes are kept close together compared to the third potential/ current electrode. It is called a lateral or gradient sonde. Some- times it is termed as paired electrode set up.

Normally the third electrode is kept at least five times the distance away from the other two closely spaced electrodes. Depending on the number of current electrodes on the sonde, it is also known as a unipolar or bipolar sonde.

For normal/potential sonde, the distance AM is called the length of the sonde and for lateral/ gradient sonde; the distance AO is the length of the sonde, where 'o' is the mid point of the paired electrodes.

Conventional lengths of the sounde are AM=16 and 64 and AO = 18.8. AM=16; is called as the short normal and AM = 64 is called as the long normal. The lengths of the sondes can be altered.

Curve Characteristics

The shape of resistivity curve depends on the type of sonde (potential/gradient), and the ratio of the following:

- The formation resistivity to the resistivity of the surrounding formation (Rt/Rs).

- The resistivity of the formation to the borehole mud resistivity (R_t/R_m).

- The length of the sonde to the borehole diameter (L/d).

- The thickness of the formation to diameter of the borehole (h/d).

Thick beds are those whose thickness is larger compared to the length of the sonde and thin beds are the ones wherein the length of the sonde is larger compared to its thickness.

Apparent resistivity curves for resistive and conductive formations and for thick and thin beds are different for potential and gradient sondes.

Normal Potential Sonde

The shape and amplitude of the curves depend on the resistivity contrast and the thickness of the target formation. In case of resistive formation of finite thickness (i.e., length of the sonde very small) the bed boundaries are determined by adding half the sonde length on either side of the borehole inflection point and the resistive bed appears one spacing length thinner than the actual thickness of the bed. The shape of the curve is symmetrical to the center of the bed boundaries.

If the formation is thin compared to the spacing of the sonde used and resistive, the curve appears as if the bed is conductive and much thicker. On either side of the bed, we have symmetrical resistive peaks. The boundaries are located by reducing half the electrode length from these peaks. In practice the peaks are poorly recorded and hence demarcation of thin formations becomes difficult.

If the target formation is by conductive nature and of finite thickness, the actual bed boundaries are located by subtracting half the spacing of the sonde on either side of the inflection point. Thus, conductive beds always appear thicker by one electrode spacing than their actual thickness.

Lateral/Gradient Sonde

The curve shapes for the gradient sonde are asymmetrical and their shapes depend on the position of the in the borehole current electrode. Two types of gradient sonde are possible, the top sensing gradient sonde and the bottom sensing gradient sonde. When the bed thickness is large compared to the length of the sonde and the formation is resistive, the gradient curves have the following characteristics in case of a bottom sensing gradient sonde.

At the upper boundary, the resistivity recorded has a minimum and the value of this minimum is less than the resistivity of the adjacent bed.

Opposite the bed, the apparent resistivity values increases from top to the bottom of the bed. This increase is small in the beginning (up to the length of the sonde (AO) from the top of the bed) and then as the inhole electrode passes through the center of the bed, the increase resistivity value is very sharp.

At the lower boundary, the resistivity curve has a maximum and below the boundary the resistivity value falls down sharply and attains a value representing the resistivity of the abject formation.

The curve characteristics will be vice-versa in the case of a top gradient sonde.

Characteristic Value of Apparent Resistivity for Beds of Finite Thickness

Apparent resistivity recorded opposite a bed varies from point to point. In practice, we use the following characteristic values of apparent resistivity.

The maximum and minimum apparent resistivity values are used for resistive and conductive beds. For a resistive bed on the potential sonde curve, maximum value is recorded against the mid-point of the bed. On the gradient log, the maximum value will be at the lower boundary in case of bottom sensing gradient sonde and at the upper boundary in case of top sensing gradient sonde.

In case of a potential sonde curve for conductive bed, the minimum value is recorded at the mid-point of the bed and for gradient sonde the minimum value is at the boundary.

The average resistivity of a bed is generally obtained by finding the area bounded by the curve and the depth axis and then dividing it by the thickness of the formation.

In practice, by first locating the upper and lower boundaries and then a straight line parallel to the depth axis is so drawn that it cuts the line representing the top and bottom boundary of the bed such that the area of triangle between the top of the bed and the curve is equal to the area bounded by the curve diagonally opposite it. In such a case the area of the rectangle so formed is equal to the area bounded by the curve. The resistivity value at the point of intersection of this line with the apparent resistivity curve represents the average resistivity value for the bed.

Optimum resistivity is the value that is close to the true resistivity of the layer. It corresponds to the value at a point on the curve approximately half the spacing of the sonde above the mid-point of the bed (top sensing gradient sonde) or half the spacing of the sonde below the midpoint of the formation (bottom sensing gradient sonde). The average resistivity value and the optimum resistivity value are usually determined for beds whose thickness is more than the length of gradient sonde.

Determination of True Resistivity of Formation

Four main factors are needed to deduce the true resistivity 'R' they are average of apparent resistivity Ra mud resistivity (R_m), the diameter of the borehole (d) and the spacing AM or AO with which R_a is measured. The value of R_a over formation is to be determined from the electric log, taking the average value where small variation is present. If the bed thickness is at least four times the spacing of the normal device or at least two times the spacing of the lateral device, R_a is used to determine the true resistivity. The value of R_m is corrected for the temperature at the depth of the formation. The borehole diameter d can be determined using caliper log drill bit diameter can be taken. The electrode spacing, AM or AO is normally shown on the log heading. In practice, normally two types of situation occur:

- Without invasion.

- With invasion in the formation of interest.

Master curves known as departure curves are used for the determination of the resistivity R_t for different levels of invaded zone.

Radio Active Logging

Geophysical methods of investigation of well section using the natural or artificially produced nuclear radiation are known as radioactive, radiation logging or nuclear logging. Gamma rays and neutrons are the two important nuclear radiations measured in well logging. These two radiations have a unique ability to penetrate high density material such as rocks, well casing and cement. Radioactive logs can be used either in cased wells or open wells. Since no direct contact with the formation is necessary, any type of bore well containing air, water or drilling mud can be logged. Electrical logging in contrast, can be done only in uncased boreholes filled with drilling mud.

Radioactive logs can be obtained from old wells where original logs were not taken. While a wide variety of nuclear logging techniques are employed in oil industry only a few of them are useful in logging water wells. They are:

- Gamma ray logging

- Gamma-Gamma logging

- Neutron logging:

  ○ Neutron-neutron logging

  ○ Neutron gamma logging

Gamma Ray Logging

Natural gamma logs are the most widely used nuclear logs in ground-water application. The most common uses are for identification of lithoglogy and for stratigraphic correlation. Gamma logs can be made with relatively inexpensive and simple equipment, and they will provide useful data under a variety of borehole conditions.

The gamma log provides a record of the total gamma radiation detected in a borehole. In water-bearing formation that are not contaminated by artificial radioisotopes, the most significant naturally occurring, gamma-emitting radioisotopes are Potassium-40 and daughter products of the uranium and thorium decay series.

Fine grained detrital sediments that contain abundant clay tend to be more radioactive than quartz, sand carbonate rocks, although numerous exceptions occur. Rocks can be characterized according to their usual gamma intensity. Limestone, and dolomite usually are less radioactive than shale; however, all these rocks can contain deposits of uranium and be quite radioactive. Basic igneous rocks usually are less radioactive a silica igneous rocks, but expectations are known. Several reasons exist for a considerable variation in the radioactivity of rocks.

The volume of material investigated by a gamma probe is related to the energy of the radiation measured, the density of the material through which that radiation must pass, and the design of the probe. Dense rock, steel casing and cement will decrease the radiation that reaches the detector, particularly from a greater distance from the borehole. Under most conditions, 90 percent of the gamma radiation detected probably originates from material within 6 to 12 inches of the borehole wall. The volume of material contributing to the measured signal may be considered approximately spherical, with no distinct boundary on the outer surface. The vertical dimension of this volume also will depend on the length of the crystal, which will affect the resolution of the thin beds. Because the detector is the center of the volume investigated, radioactivity measured when the detector is located at a bed contact will be an average of the two beds. The actual radioactivity of beds with a thickness less than twice the radius of investigation will not be recorded.

The API gamma-ray unit is defined as 1/200 of the difference in deflection of a gamma log between an interval of negligible radioactive proportions of radioisotopes as an average shale but about twice the total radioactivity. One or more filed standards is needed when calibrating in a pit or well and when calibration frequently during logging operations to assure that a gamma logging system is stable with respect to time and temperature. Field standards may consist of radioactive sources that can be held in one or more fixed position in relation to the detector while readings are made. If this approach is used, the probe is best located at least several feet above the ground and distant from a logging truck that contains other radioactive sources that could contribute to the background radiation. Radiation measurements around the logging truck will determine the proper distance.

Because of numerous deviations from the typical response of gamma logs to lithology, some background information on each new study area is needed to decrease the possibility of errors in interpretation.

In igneous rocks, gamma intensity is greater in silica rocks, such as granite, than in basic rocks. Orthoclase and biotite are two minerals that contain radioisotopes of sedimentary rocks of chemical decomposition has not been too great. Gamma logs are used widely in the petroleum industry to establish the clay or shale content of reservoir rocks; this application also is valid in coal exploration studies.

The increase in radioactivity from an increase in fine grained materials has been the basis for a number of studies relating gamma log response to permeability in various parts of the world, such as the Denver Julesberg basin in Colorado, Texas, USA, India and Russia.

Gamma – Gamma Logging

Gamma-Gamma logs (also known as density logs) are records of the intensity of gamma radiation from a source in the probe after it is back scattered and attenuated within the borehole and surrounding formation. The main uses of gamma- gamma logs are for identification of lithology and the measurement of the bulk density and porosity of rocks. These logs may also be used for locating cavities and cement outside the casing of well.

The gamma-gamma probe contains a source gamma rays, generally Cobalt- 60 or Cesium-137, shielded from sodium iodide detector by Mallory-1000 metal or lead spacers. Gamma rays from the source penetrate and are scattered and absorbed by the fluid, casing and formatting surrounding the probe. Gamma radiation is absorbed and (or) scattered by all material through which travels. Degradation of photon energy takes place by three main processes:

- The Compton scattering, in which a gamma rays less part of its energy to an orbital electron (z) and occurs with gamma photons from 01. To 1 mev.

- The photoelectric effect, in which an ejected orbital electron completely absorbs the photon energy, is proportional to $Z^{4.5}$ and occurs with photons 0.1 mev or less.

- Pair production, which occurs as the photon approaches the nucleus and completely converts itself into a pair of electrons, is proportional to Z^2 and requires gamma energy greater than 1.02 mev.

Compton scattering is probably the most significant process taking place in gamma-gamma logging. Some photoelectric absorption also takes place because of degradation of photon energy by scattering, but the effect on a log may be reduced by energy discrimination. In the Compton range the gamma radiation absorbed is proportional to the electron density of the material penetrated, but it is affected by the chemical nature of the medium. The electron density is approximately proportional to the bulk density of most materials penetrated in logging.

The Gamma-gamma log can also be used to identify borehole enlargements through casing. The water level and significant change in fluid density will also apparent on gamma-gamma logs.

In gamma-gamma logging the relative percentage of gamma photons absorbed and scattered depends to a large degree upon the type and size of the source, spacing between the source and detector, and the borehole diameter. The radius of investigation depends on these same factors in addition to the bulk density. In general source strength large diameter holes.

Modern gamma-gamma probes are decentralized and side collimated. Side collimation with heavy metal tends to focus the radiation from the source and to limit the detected radiation to that part of the wall of the borehole in contact with the source and detectors. The decentralizing caliper arm also provides a log of hole diameter. The decentralized tool has the advantage of being much less affected by changes in drill-bit size, sonde position in the hole, or density of fluid in the hole.

The gamma-gamma loggers can be calibrated in terms of density by comparing the probe responses against known formation with densities of core samples from the same spot. Measurements of two to three different formations, with their densities covering the density range of interest, are required for constructing the calibration graph in semi-log scale. The gamma gamma logs can be used to determine the formation porosity using the relation:

Porosity = grain density – Bulk Density (from log) /Grain density-fluid density

Grain density can be derived from laboratory analysis of cores or cuttings. The fluid density in most water wells may be assumed to be l gm/cc. If the fluid is highly saline, laboratory measurement of density may be necessary. If the same lithologic unit is present below and above the water table, or if gamma-gamma measurement can be made after the drawdown, it should be possible to derive specific yield from gamma-gamma log specific yield should be proportional to the difference between the bulk density of saturated and drained sediments, if the porosity, and grain density are not changed. Bulk density may be read directly from a calibrated and corrected log or derived from a chart providing correction factors. Errors in bulk density obtained by gamma-gamma methods are of the order of +/- 0.03-0.04 gm/cc. Errors in porosity calculated from log bulk densities depend on the accuracy of grain and fluid densities used. For certain source to detector spacing's and over a limited density range, a linear relationship is obtained when bulk density is plotted against the logarithm of count rate. In addition to determining porosity, the gamma-gamma log may be used to locate casing or collars or the position or the position of grout outside of the casing.

Neutron Logging

In neutron logging a neutron source is lowered along with the detectors into the borehole. The source is fixed at the end of the probe and above it the counters are placed. The spacing between the neutron source and counters may be 12 to 27". Depending upon the recording capability of the counters used in the neutron probe, three different types of neutron logs are possible. These are the neutron gamma log, the neutron thermal neutron log and the neutron epithermal neutron log. The neutron logs are chiefly used for the measurement of moisture content above the water table and porosity below the water table. In most of the modern logging equipments Americium-Beryllium source 106^6 neutrons per second and the half life of the source is 458 years. So the decay correction factor is very small the source emits fast neutron emits fast neutrons having an average energy level of the order of 10^5ev.

Various types of neutron logs are made by counting neutrons present at different energy levels the neutron thermal neutron probe responds chiefly to thermal neutrons i.e., the neutron having energy level between 0.025 Mev and 0.1 Mev and the neutron epithermal neutron tool responds mostly to the neutrons having energy level between 0.1 ev and 100ev. The neutron-neutron logs do not have the problem of shielding the counters from the gamma radiation of the source since the counters used in this type of logging system are not sensitive to gamma radiation. In the neutron thermal neutron log, the neutrons are determined with a proportional counter after they reach thermal energy level. While in the case of neutron epithermal neutron log the counter measures the neutrons just before they reach the thermal level. The counters used in this type of instruments are scintillation counters. The neutron thermal neutron log is sensitive to the variation in the capture cross section of the formation elements while the neutron epithermal neutron log is completely insensitive to variation in capture cross section because it measures the neutron at an energy level before the reaction take place.

Radius of Investigation

The radius of investigation of neutron device is from 6 inches for high porosity saturated formation to 2 feet for low porosity or dry rocks. The neutron logs are the most useful techniques as applied to ground water investigations, because most of the probe response is due to hydrogen and therefore water use:

- The neutron logs are chiefly used for the measurement of moisture content above the water table and total porosity below the water table.

- The combined study of thermal neutron log and natural gamma ray log helps in differentiating the formations from each other. The gypsum and anhydrite both emit very low intensity gamma radiation and on the logs they will have the same counting level which makes it difficult to interpret the logs. But if the thermal neutron log of the same borehole is available it is possible to differentiate them because the anhydrite gives low counting rate while the gypsum gives high counting rate. Thus comparison of these logs helps in making quantitative interpretation of the log.

- The epithermal neutron logs provide the highest percentage of response due to hydrogen and are least affected by the chemical composition of rocks and the fluids they contain.

- The neutron gamma ray log is very sensitive to chloride content, while the neutron epithermal neutron log is insensitive to the chemical composition of fluids. This comparison of these two logs reveals the presence of chloride in the formation. In petroleum logging the direct comparison of neutron-neutron logs with neutron gamma logs gives the idea about saline formation water because the increase in count rate on the neutron gamma ray log is supposed to be due to chlorine and thus due to NaCl concentration. described how neutron logging devices can be used to determine the specific yield of unconfined aquifers.

- A conventional pumping test neutron logging methods were used simultaneously to determine the specific yield of the aquifer in the alluvium formation. Before the pumping test was stated the moisture percentage by volume was determined with neutron log for each foot of depth of the saturated zone. The pumping test was then carried out and the amount of water pumped out was then measured. Then again the neutron log was run and the moisture percentage was determined. The specific yields calculated by pumping test data and by neutron logging were very close to each other.

Caliper Logging

Caliper logs provide a continuous record of borehole diameter and are used extensively for ground water application. Changes in borehole diameters may be related to both drilling technique and lithology. Caliper logs are essential to guide the interpretation of other logs, because most of them are affected by changes in borehole diameter. Caliper logs also are useful in providing information on well contraction, lithology, and secondary porosity. Many different types of caliper logs are described in detail by. Single arm caliper probes commonly are used to provide a record of borehole diameter, while running another type of log. The single arm also may be used to decentralize a probe, such as a side collimated, gamma-gamma probe. This probe has advantages in this the arm generally follows the high side of a deviated hole. A three arm averaging caliper probe does not function properly in highly deviated boreholes, because the weight of the tool forces on arm to close, which closes the other two arms.

Calibration of caliper probes is done most accurately in rings of different diameters. Because large cylinders occupy considerable space in a logging truck, it is common practice to use a metal plate for on-site calibration. The plate is drilled and marked every inch or two and machined to fit over the body of the probe, one arm is placed in the appropriate holes in the range to be logged; the pen location is labeled on the analog chart and a digital value is recorded, if applicable.

Heavy drilling mud will prevent caliper arms from opening fully and thick mud cake may prevent accurate measurement of drilling diameter.

A Caliper log is needed to interpret many other logs, it needs to be made before the casing is installed in borehole that is in danger of caving. When borehole conditions are questionable, the first log made generally is the single point resistance logs, because it will provide some lithology information; the probe is relatively in- expensive, if it is lost. If no serious caving problems are detected during the running of the single point resistance log, a caliper log needs to be run before the casing is installed so it can be used to aid the analysis nuclear logs made through the casing. Data for extremely rough intervals of the borehole wall, with changes in diameter of several inches, cannot be corrected based on caliper logs; data for these intervals need to be eliminated from quantitative analysis.

Caliper logs can provide information on lithology and secondary porosity. Hard rocks like limestone will have a smaller diameter than adjacent shales. Thin beds may result in an irregular trace. Secondary porosity, such as fractures and solution openings, may be obvious on a caliper log. Caliper logs have been used to correlate major producing aquifers in the Snake River plain in Idaho. Vesicular and Scoriaceious tops of basalt flows, cinder beds, and caving sediments were identified with three arm caliper logs. In the basalt of the quaternary snake river group, caliper logs also were used to locate the optimum depth for cementing and to estimate the volume of cement that might be required to fill the annals to a preselected depth.

Similarly a caliper log is used to calculate the volume of gravel pack needed and to determine the size of casing that can be set to select the depth.

Temperature Logging

Temperature logs can provide useful information on the movement of water through a borehole, including the location of depth intervals that produce or accept water. Thus, they can provide information related to permeability distribution and relative hydraulic head. Temperature logs can be used to trace the movement of injected water or waste, to locate the cement behind casing and to correct other logs that are sensitive to temperature. Though the temperature sensor only responds to the temperature of the water or air in the immediate vicinity, recorded temperatures may indicate the temperature of adjacent formation and their contained fluids. Formation temperature may be indicated if no flow exists in the borehole and if equilibrium exists between the temperature of the fluid and the temperature of the adjacent rocks.

The differential temperature log can be considered the first derivative of the temperature log; it can be obtained by two different types of logging probes or by computer calculation from a temperature log. Logging speed needs to be maintained accurately for this method departure from the reference gradient will be recorded as deflections on the log.

Temperature logs can aid in the solution of a number of ground water problems if they are properly run under suitable conditions and if interpretation is not over simplified. If there is no flow in or adjacent to borehole, the temperature gradually will increase with depth, as a function of the geothermal gradient. Typical geothermal gradients range between 0.47 to 0.6°C/100 ft of depth; they are related to the thermal conductivity or thermal resistivity of the rocks adjacent to the borehole and the geothermal heat flow from below. Conway developed a computer program for correcting digitized temperature data, calculating temperature gradient or differential temperature logs.

Temperature data from wells are also used to calculate water density, viscosity, and thermal conductivity and to develop heart flow maps, which can be used to estimate the fluid flux, particularly, in geopressured aquifers.

Chapter 4

Petroleum Extraction

The process through which usable petroleum is drawn out from beneath the surface of the Earth is called petroleum extraction. This process is divided into a few phases, namely, field appraisal phase, field development phase, petroleum production phase and well abandonment phase. This chapter discusses in detail these phases of petroleum extraction.

After successful drilling exploration wells, the appraisal stage of the lifecycle starts. The main purpose of this phase is to improve the field description through further data acquisition and to reduce the uncertainty or possibility of losses about the size, shape and marketability of the oil and gas reservoir.

Oil extraction or petroleum extraction is the extraction of usable crude oil from regions having suitable surface and underground configurations.

Economics of Oil Extraction

The primary challenge in oil extraction is to find ways to minimize upfront costs and optimize the extraction process to yield maximum profitability. Establishing an oil extraction system should be both technically feasible and economically viable.

Major factors that affect oil prices by impacting the extraction process and cost are:

- The subsurface condition of a reservoir: Highly feasible extraction sites have already been explored. The cost of extracting from the remaining sites is higher due to their less favorable geographic conditions.

- Drilling procedure: Onshore drilling generally requires low investments and bear lower economic risks. Offshore drilling consumes over 65% of the planned investments before the extraction process commences, increasing the risk of losses.

- 'Peak oil' concerns: This implies the point at which oil production begins to decline due to the maximum oil extraction rate being attained. Concerns over reaching peak oil have escalated the oil extraction costs.

Oil Extraction: Alternative Methods

Some alternative methods of oil extraction are:

- Fischer-Tropsch Method: It includes a catalyzed chemical reaction of converting methane and coal into various forms of liquid hydrocarbons. Cobalt and iron are the commonly used catalysts in the method. This process helps to produce synthetic lubricants (synthetic petrol) to power automobiles and jets. The technology has been commercialized in Malaysia (Shell) and South Africa (Sasol).

- Karrick Process: The process was developed by Lewis C. Karrick, an oil-shale technologist. It involves low-temperature carbonization (LTC) of carbonaceous materials, such as lignite, coal and shale. These are shielded from air exposure and heated at a temperature between 680°F (360°C) and 1380°F (749°C). This helps to extract oil and gasoline for commercial use.

- Thermal Depolymerization (TDP): The process starts with hydrous pyrolysis, which involves the heating of organic compounds at a high temperature (in the presence of water). This helps in reducing complex organic materials, such as biomass, into light crude oil.

Well Drilling

Drilling is a process whereby a hole is bored using a drill bit to create a well for oil and natural gas production. There are various kinds of oil wells with different functions:

- Exploration wells (or wildcat wells) are drilled for exploration purposes in new areas. The location of the exploration well is determined by geologists.

- Appraisal wells are those drilled to assess the characteristics of a proven petroleum reserve such as flow rate.

- Development or production wells are drilled for the production of oil or gas in fields of proven economic and recoverable oil or gas reserves.

- Relief wells are drilled to stop the flow from a reservoir when a production well has experienced a blowout.

- An injection well is drilled to enable petroleum engineers to inject steam, carbon dioxide and other substances into an oil producing unit so as to maintain reservoir pressure or to lower the viscosity of the oil, allowing it to flow into a nearby well.

The process of drilling an oil and natural gas production well involves several important steps:

- Boring: A drill bit and pipe are used to create a hole vertically into the ground. Sometimes, drilling operations cannot be completed directly above an oil or gas reservoir, for example, when reserves are situated under residential areas. Fortunately, a process called directional drilling can be done to bore a well at an angle. This process is done by boring a vertical well and then angling it towards the reservoir.

- Circulation: Drilling mud is circulated into the hole and back to the surface for various functions including the removal of rock cuttings from the hole and the maintenance of working temperatures and pressures.

- Casing: Once the hole is at the desired depth, the well requires a cement casing to prevent collapse.

- Completion: After a well has been cased, it needs to be readied for production. Small holes called perforations are made in the portion of the casing which passed through the production zone, to provide a path for the oil or gas to flow.

- Production: This is the phase of the well's life where it actually produces oil and/or gas.

- Abandonment: When a well has reached the end of its useful life (this is usually determined by economics), it is plugged and abandoned to protect the surrounding environment.

Drilling is a relatively well-understood technological process but no two wells are the same and therefore risk management is important. The largest mainstream concern with drilling is the risk of blowouts, which is the uncontrolled release of oil and natural gas from a well due to issues with pressure management. With modern technology blowouts are preventable. However a high level of diligence is required by operators and regulators to ensure this does not happen.

In addition to this, there are a wide array of drilling activities that can cause adverse environmental impacts. For example, ground clearing can have adverse effects on the ecological surroundings. Air quality and waste management from construction and during drilling can be an issue. The increase and vehicle and pedestrian traffic also creates an impact on the local environment.

Rotary Drilling Rig Components

During the first phase of the development of the well, a rotary drilling rig is installed to bore a hole in the ground and reach the oil reservoir. The main rotary drilling rig components are derrick or mast, power and prime movers, hoisting equipment, rotating component, circulating system, tubular and tubular handling equipment and bit.

Derrick is mainly used offshore and is a large load-bearing vertical structure, usually of bolted construction and pyramidal in shape, for the equipment used to lower and raise the drill string into and out of the wellbore. The height of the derrick does not affect its load-bearing capacity, but it shows the maximum length of the drill pipe section. The standard derrick has square-shaped rig floor with four legs standing at the corners of the substructure. It provides work space for the necessary equipment on the rig floor.

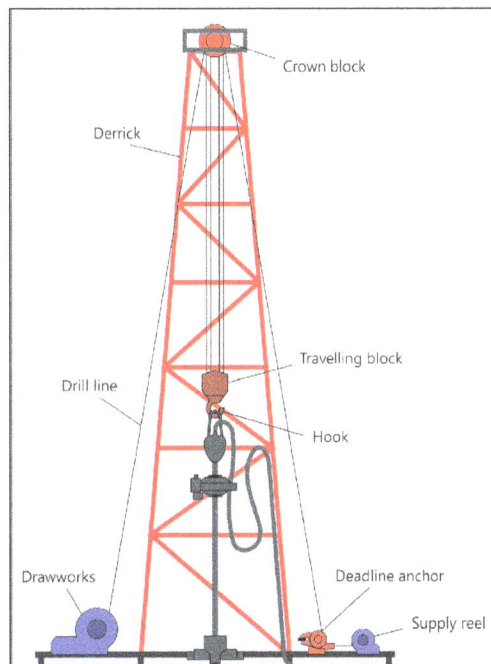

Mast is mainly used with onshore rigs and is a portable derrick that can be raised as unit but for the transporting can be divided into two or more sections. It is usually rectangular or trapezoidal in shape.

Power and prime movers: The power generated by the power system is used for five main operations such as rotating, hosting, drilling fluid circulation, rig lighting and hydraulic systems. It is important to note that the most of the generated power is consumed by the hoisting and drilling fluid circulation systems. Internal combustion engine (mostly diesel) connected to electric generators or turbine is the source of power on the rig. Some rotary rigs can use electricity directly from power lines.

Hoisting component is used to perform all lifting activities on the rig and helps in lowering or raising equipment into or out of the well. It consists of drawworks, crown block, traveling block, deadline anchor, supply reel and drilling line.

- Drawworks is the main operating component of the hoisting system and is used to transmit power from prime movers to the hoisting drum that lifts drill string, casing or tubing string out of and to lower it back into the borehole. They consist of a large diameter steel spool, brakes, a power source and assorted auxiliary devices. The primary function of the drawworks is to reel out and reel in the drill line, a large diameter wire rope, in a controlled manner. The speeds for hoisting the drill string could be changes by driller via integrated gear system.

- Crown block is fixed assembly of sheaves (single or double) with a wire rope drilling line running between it and is located at the top of the derrick or mast and over which the drilling line is threaded. It is used to change the direction of pull from the drawworks to the traveling block.

- Traveling block and hook combination is used to safely and efficiently raise or lower tools and equipment in the well. It is the set of sheaves or pulleys through which the drill line (wire rope) is threaded or reeved, is opposite the crown block and enabling heavy loads to be lifted out of or lowered into the wellbore. Hook is located beneath the traveling block and is used to pick up and secure the swivel and Kelly.

- Deadline anchor is usually bolted on to the substructure and is the equipment that holds down the deadline part of the wire rope. It provides weight measurements and secure deadlines.

- Supply reel is a spool that stores the unused portion of the drill line.

- Drill line is the wire rope used to support the drilling tools. It is threaded or reeved through the traveling block and crown block to facilitate the lowering and lifting of the drill string into and out of the borehole. Drill line then clamped to the rig floor by the deadline anchor.

Rotating component is the equipment responsible for rotating the drill string. It consists of the swivel, Kelly spinner, Kelly or top drive, Kelly bushing, master bushing and rotary table.

- Swivel is a mechanical device that is hung from the hook of the traveling block to support the weight of the drill string and allows it to rotate freely. It provides connection for the rotary hose as well as passageway for the flow of drilling fluid into the drill stem.

- Kelly spinner is a pneumatically controlled device mounted below the swivel to spin the Kelly to make up tool joints when making connections.

- Kelly is the heavy steel square or hexagonal member that is suspended from the swivel through the rotary table and connected to the topmost joint of drill pipe to turn the drill stem as the rotary table turns. It has a hole drilled through the middle that permits fluid to be circulated into the drill stem and up the annulus or vice versa. The Kelly goes through the Kelly bushing, which is driven by the rotary table.

- Top drive is a hydraulically powered device on the drilling rig and is located at the swivel place. It allows the drill stem to spin and facilitate the process of drilling a borehole. Top drive means a power swivel, which directly turns the drill string without need for a Kelly and rotary table.

- Kelly bushing is a device that fits into a part of rotary table called master bushing, transmits torque to the Kelly and simultaneously permits vertical movement of the Kelly to make hole. The Kelly bushing as Kelly is square or hexagonal and has an inside profile matching the Kelly's outside profile with slightly larger dimensions so that the Kelly can freely move up and down inside.

- Master bushing is a tool that fits into the rotary table of a drilling rig to accommodate the slips and drive the Kelly bushing so that the rotating motion of the rotary table can be transmitted to the Kelly.

- Rotary table is section of the drill floor used to turn the drill stem. It has a beveled gear arrangement to create the rotational motion and opening into which bushings are fitted to drive and support the drilling assembly.

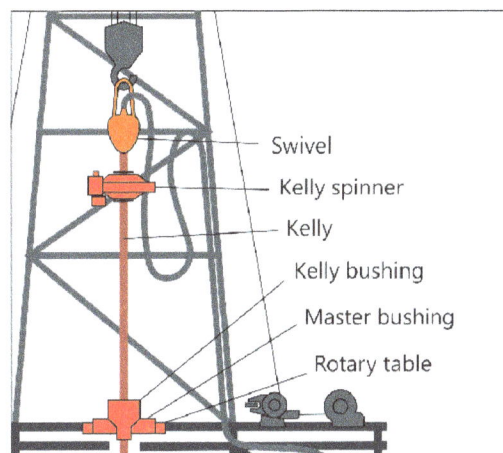

Rotating equipment of the drilling rig.

Circulating component is the rig equipment responsible for the movement of drilling fluid within the well as well as solids removal incurred by the drilling fluid. Normally, the circulation would start from the mud pits or tanks that are located besides the rig. Powerful pumps force the drilling through the surface high-pressure connections to a set of valves called pump manifold, located at the derrick floor. Then, the fluid goes up the rig within a pipe called standpipe to approximately 1/3 of the height of the mast. From there, the drilling fluid flows through a flexible high-pressure rotary hose to the top of the drill string. The flexible hose allows the fluid to flow continuously as the drill string moves up and down during normal drilling operations. The fluid enters in the drill string through a special piece of equipment called swivel located at the top of the Kelly. The swivel permits rotating the drill string while the fluid is pumped through the drill string. In wellbore, the drilling fluid then floes down the rotating string and jets out through the nozzles in the drill bit at the bottom of the hole. Drilling fluid carrying the drilled cuttings and flows out the center of the drill bit and is forced back up the outside of the drill pipe between the drill string and walls of the well (annular) onto the surface of the ground where it is cleaned and circulated back to the well. The cleaning process starts from the shale shaker, which is basically a vibrating screen.

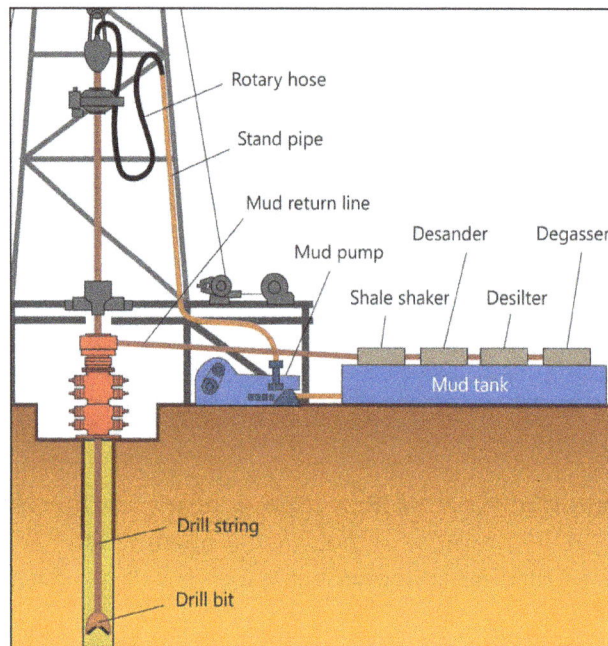

Circulation system of the drilling rig.

This will remove the larger particles, while allowing the residue to pass into settling tanks. The finer material can be removed using other solids removal equipments such as desander and desilter. If the mud contains gas from the formation, it will be passed through a degasser that separates the gas from the liquid mud. Having passed through all the mud processing equipment, the mud is returned to the mud pits or tanks for recycling.

The principal components of the drilling fluid circulation system are as follows:

- Mud pump is a large, high-pressure and high-volume pump used to circulate the drilling fluid down the drill pipe and out of the annulus on an oil rig. It could be double acting duplex (2 cylinder) pump, which has four pumping actions per pump cycle or single acting triplex (3 cylinder) with three pumping actions per pump cycle whose pistons or plungers travel in replaceable liners and are driven by a crankshaft actuated by an engine or motor.

- Pump manifold is an arrangement of piping and valves that receives drilling fluid from mud pumps and transmits the drilling fluid to the succeeding circulating component. It is designed to control, distribute and monitor drilling fluid flow.

- Stand pipe is the vertical rigid pipe rising along the side of the derrick or mast, which joins mud pump manifold to the rotary hose.

- Drill string is the mechanical assemblage connecting the rotary drive on the surface to the drilling bit on bottom of the wellbore.

- Mud return line or flow line is the large diameter metal pipe and is the passageway of the drilling fluid as it comes out of the well.

- Shale shaker is the primary solids-removing device with one or more vibrating screens, which is used to remove cuttings from the circulating fluid for reuse. Screens vibrate while the mud flows on top of it. The liquid phase with solids which are smaller than the wire mesh pass through the screen as well as larger solids are retained on the screen and eventually fall to the special container and can be disposed in an environmentally friendly manner.

- Desander is a centrifugal device for removing sand-size particles from the drilling fluid to prevent abrasion of the pumps. There are no moving parts of a desander, and the removal of particles is done by gravity and pressure. As the drilling fluid flows around and gradually down the inside of the cone shape, particles are separated from the liquid by centrifugal forces.

- Desilter is also a centrifugal device for removing free particles of silt from the drilling fluid. Comparing with desander, its design incorporates a greater number of smaller cones, which allow removing smaller diameter particles than a desander does.

- Degasser is device designed to remove air, methane, hydrogen sulfide (H_2S), carbon dioxide (CO_2) and other gases from drilling fluids and allow it to be reused continuously. It helps to reduce the risk of explosions and other dangers during the drilling process.

- Mud pit is an excavated earthen-walled pit and is used only to store used or waste drilling fluid and cuttings.

- Mud tank is an open-top steel container with possibility to observe the consistency of drilling fluid and monitor it level in the tanks. It is used as a reserve store for the drilling fluid.

Tubular and tubular handling equipment. Tubular consists of the following equipments:

- Drill pipe is the longest section of the drill string and is heavy hot-rolled, pierced and seamlessly tubing. It connects the surface equipment with the bottom hole assembly and the bit is used to rotate the bit and for drilling fluid circulation.

- Drill collar is thick-walled, heavy and large diameter steel tube placed between the drill pipe and the bit in the drill stem to provide weight on a bit. It can be cylindrical or spiral shape and is threaded at both ends (male and female) to allow multiple drill collars to be joined above the bit assembly.

- Heavy weight drill pipe is thick-walled tube and is used as transition pipe between drill collar and drill pipe. In high-angled and horizontal wellbore, it is used in lieu of drill collers.

- Subs are short component of the drill string, threaded piece of pipe used to adapt parts of the drilling string that cannot otherwise be screwed together because of difference in thread size or design.

Tubular handling equipment is made of the following equipments:

- Elevator clamps that grip a stand of casing, tubing, drill pipe or drill collars so that the stand or joint can be lifted and lowered into the wellbore opening of the rotary table. The elevators are connected to the traveling block by means of bails, which are solid steel bars with eyes at both sides. Elevator could be side door, center latch or single joint types.

- Elevator links is device designed to support the elevators and attach them to the hook.

- Slips are a wedge-shaped piece of metal with teeth or other gripping elements that supports and transmits the weight of the drill string to the rotary table and are used to hold the pipe in place as well as to prevent pipe from slipping down into borehole. Different types of slips are used during oil well drilling such as drill pipe, drill collar or casing slips.

- Safety clamp is a mechanical device used on tubulars above the slips and is used to keep parts of the tool string from falling down the wellbore if other safety measures fail.

- Tongs are large wrenches used to make or break out tubular. It must be used in opposing pairs—make up or breakout tongs to make or break connection.

- Drill pipe spinner is a pneumatically operated device usually suspended on the rig floor used to make fast connections and spin off of drill pipe.

- Iron roughneck is a pneumatically operated machine that replaces the functions performed by the Kelly spinner, drill pipe spinner and tongs and is used to connect and disconnect tubular.

Drill bits are cutting tools used to create cylindrical holes. Bits are located at the bottom of the drill string and are suited for particular conditions, such as formation, which is to be drilled. There are three different types of bit designs, such as:

- Roller cone bits with milled tooth or tungsten carbide insert (TCI) could have 2–6 cone-shaped steel devices that are free to turn as the bit rotates.

- Fixed cutter bits could be drill bit or core bit. The first one could be polycrystalline diamond compact bit (PDC-bit), surface set diamond bit and impregnated diamond bit. It consists of bit bodies and cutting elements integrated with the bit bodies and do not have moving parts.

- Hybrid bits combine both rolling cutter and fixed cutter elements.

If the drill bit needs to be changed, the whole string of pipe must be raised to the surface.

Classification of the Drilling Fluids

Modern drilling fluids (muds) are complex heterogeneous fluids (water based, oil based) and are complex mixtures of more than 200 minerals and chemicals. It is used in a drilling operation and circulates from the surface, down the drill string, through the bit and back to the surface via the annulus. The original use of the drilling fluids was to remove cuttings continuously. Progress in drilling engineering demanded more sophistication from the drilling mud. In order to enhance the usage of rilling fluids, numerous additives were introduced and a simple fluid became a complicated mixture of liquids, solids and chemicals. As the drilling fluids evolved, their design changed to have common characteristic features that aid in safe, economic and satisfactory completion of a well. In addition, drilling fluids are also now required to perform following functions:

- Clean the rock formation beneath the bit for rock cuttings.

- Remove cutting from the well.

- Control formation pressures while drilling and maintain wellbore stability.

- Suspend and release cuttings.

- Seal permeable formations to prevent excessive mud loss.

- Minimize reservoir damage by using reservoir drill-in fluid.

- Cool, lubricate and clean the bit and drilling assembly.

- Transmit hydraulic energy to downhole assembly.

- Ensure adequate formation evaluation.

- Control corrosion.

- Facilitate downhole measurement (measurement while drilling, logging while drilling).

- Facilitate cementing and completion.

- Minimize impact on the environment.

However, excessive use of oil-based drilling fluids may harm the environment and it is important to develop more environmentally friendly drilling fluids. In this respect, water-based drilling fluids are more acceptable. As well known, bentonite is widely applied in the water-based drilling fluids, which could enhance the clean properties and form a thin filter with low permeability. The functions of bentonite are to make the fluids more viscous and reduce the loss of fluids. There are four types of drilling fluids:

Water-based drilling fluid (WBM) is the mud in which water is continuous phase. The water could be fresh, brackish or seawater. The most basic WBM system begins with water, then clays and other chemical and is incorporated into the water to create a homogenous blend. The clay (called "shale" in its rock form or bentonite) is frequently referred to in the oilfield as "gel." Many other chemicals (e.g. potassium formate, $KHCO_2$) are added to a WBM system to achieve various effects, including velocity control, shale stability, enhance drilling rate of penetration, cooling and lubricating of equipment.

Advantages:

- Low cost.

- High rate of penetration.

- Good cuttings removal.

- Good geoscientific investigations.

- The pressure in the cutting area increases with increasing hydrostatic pressure of drilling fluid.

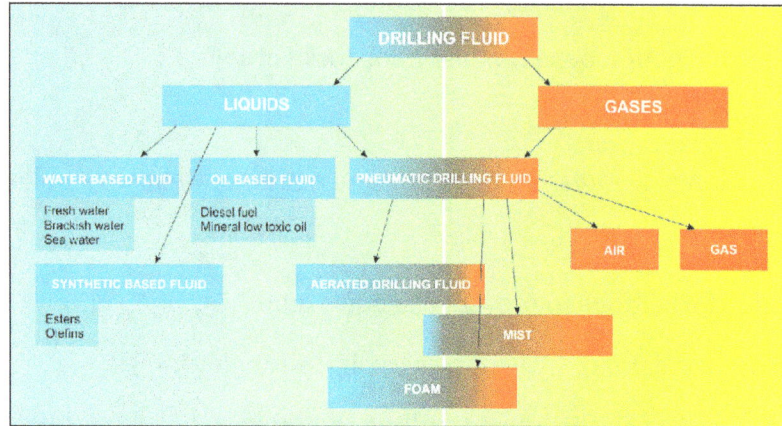

Classification of the drilling fluids.

Disadvantages:

- Low borehole stability.

- Insufficient cutting transport efficiency.

- Insufficient lubricating properties.

- Drilling fluid loss.

Oil-based drilling fluid has best technical properties such as stability, lubricity and temperature stability. Oil-based mud can be a mud where the base fluid is a petroleum product such as diesel fuel or mineral low toxic oil. The authorities do not permit the discharge of oil-based drilling fluid and cuttings drilled with oil-based drilling fluids because of their special nature of being a mixture of two immiscible liquids (oil and water). In that case special treatment and testing are required. The terms oil-based mud and inverted or invertemulsion mud used to distinguish among the different types of oil-based drilling fluids. Traditionally, an oil-based mud is a fluid with 0–5% by volume of water, whereas an invertemulsion mud contains more than 5% by volume of water.

Advantages:

- Excellent lubricating properties (reduce drilling torque and drag).

- Good temperature stability.

- Favorable to borehole stability.

- High rate of penetration.

- Will not hydrate clays.

- Long bit life.

- Low reservoir damage.
- Low drilling fluid loss.
- Salt not dissolved.
- Corrosion resistance.
- Can be reused.

Disadvantages:

- High initial cost.
- Electric log difficulty.
- Viscosity varies with temperature.
- Environmental issue.
- Difficult to keep the rig clean while drilling.
- Difficult to identify gas kick.
- Messy working environment.
- Fire hazards.

Synthetic drilling fluids are based on ether, ester or olefin. They have technical properties that are similar to oil-based drilling fluids and are most often used on offshore rigs or in environmentally sensitive areas, because it has the properties of an oil-based mud, but the toxicity of the fluid fumes is much less than an oil-based fluid. This is often used on offshore rigs.

Advantages:

- Favorable to borehole stability.
- High rate of penetration.
- Good wellbore stability.
- Good control of drilling fluid properties.
- Good cutting transport efficiency and removal.
- Good filtration properties.

Disadvantages:

- Complex system with high solid content.
- Geoscientific investigations difficulty.

Pneumatic drilling fluids—Fluids, which are based on air/gas, mist, aerated fluid or foam. Air drilling is used primarily in hard rock areas and in special cases to prevent formation damage while drilling into production zones or to circumvent severe lost-circulation prob- lems. Air drilling includes dry air drilling, mist or foam drilling and aerated mud drilling. In dry air drilling, dry air/gas is injected into the standpipe at a volume and rate sufficient to achieve the annular velocities needed to clean the hole of cuttings. Mist drilling is used when water or oil sands are encountered that produce more fluid than can be dried up using dry air drilling. A mixture of foaming agent and water injected into the air stream, producing foam that separates the cuttings and helps remove fluid from the borehole. In aerated mud drilling, both mud and air pumped into the standpipe at the same time. Aerated mud is used when it is impossible to drill with air alone because of water sands and/ or lost-circulation situations.

Advantages:

- High rate of penetration.
- Low reservoir damage.
- Good bit performance.
- Low drilling fluid loss.
- Low water consumption.
- Low air quality requirements for foam drilling.
- Low hydrostatic pressure.
- Good cleaning of the borehole.

Disadvantages:

- There are restrictions on the possible lithological structures.
- Drilling could be limited by the length of the horizontal section of the well.
- Possibility of fire.
- Possible additional costs to rent equipment.
- Gas costs.
- Gas and foam utilization issues.
- Aerated fluids require specialized equipment for the injections.
- Aerated fluids and foam have potential corrosion problems and the need to use additional inhibitor.
- The quality of the foam changes in exchange pressure.
- The foam is a complicated system and may require computer modeling of foam movement in the borehole.

Mud Ingredients

Various materials may be added at the surface to change or modify the characteristics of the mud:

- Weighting materials (usually barite) are added to increase the density of the mud, which helps to control subsurface pressures and build the filter cake. Salts are sometimes added to protect downhole formations or to protect the mud against future contamination, as well as to increase density. Dispersants or deflocculants may be added to thin the mud, which helps to reduce surge, swab and circulating pressure problems.

- Viscosifying materials (clays, polymers and emulsified liquids) are added to thicken the mud and increase its hole cleaning ability.

- Filtration control materials. Clays, polymers, starches, dispersants and asphaltic materials may be added to reduce filtration of the mud through the borehole wall. This reduces formation damage, differential sticking and problems in log interpretation.

- pH control and lubricating materials. Mud additives may include lubricants, corrosion inhibitors, chemicals that tie up calcium ions and flocculants to aid in the removal of cuttings at the surface. Caustic soda is often added to increase the pH of the mud, which improves the performance of dispersants and reduces corrosion.

- Other additives. Preservatives, bactericides, emulsifiers and temperature extenders may all be added to make other additives work better.

Most of these additives have distinct properties that help in countering specific challenges encountered during the drilling process as well as in accomplishing the drilling work with efficiency and precision. However, to select the proper fluid, it is necessary to calculate the cost of the fluid, understand the environmental impact of using the fluid and to know the impact of the fluid on production from the pay zone.

Field Appraisal Phase

The exploration phase of the petroleum field life cycle closely links with the next stage which is known as the appraisal phase.

Once an exploration well has found hydrocarbons, considerable effort will still be required to accurately assess the potential of the discovery and the role of appraisal is to provide cost-effective information that will be used for subsequent decisions (development).

During appraisal, more wells are drilled to collect information and samples from the reservoir and other seismic survey might also be acquired in order to better delineate the reservoir.

This phase of the E and P process aims to:

- Reduce the range of uncertainty in the volumes of hydrocarbons in place.

- Define the size and configuration of the reservoir.

- Collect data for the prediction of the performance of the reservoir during the forecasted production life.

Having defined and gathered data adequate for an initial reserves estimation, the next step is to look at the various options to develop the field. During the appraisal phase, reservoir engineering increases it's contribute in reaching the technical and economic targets.

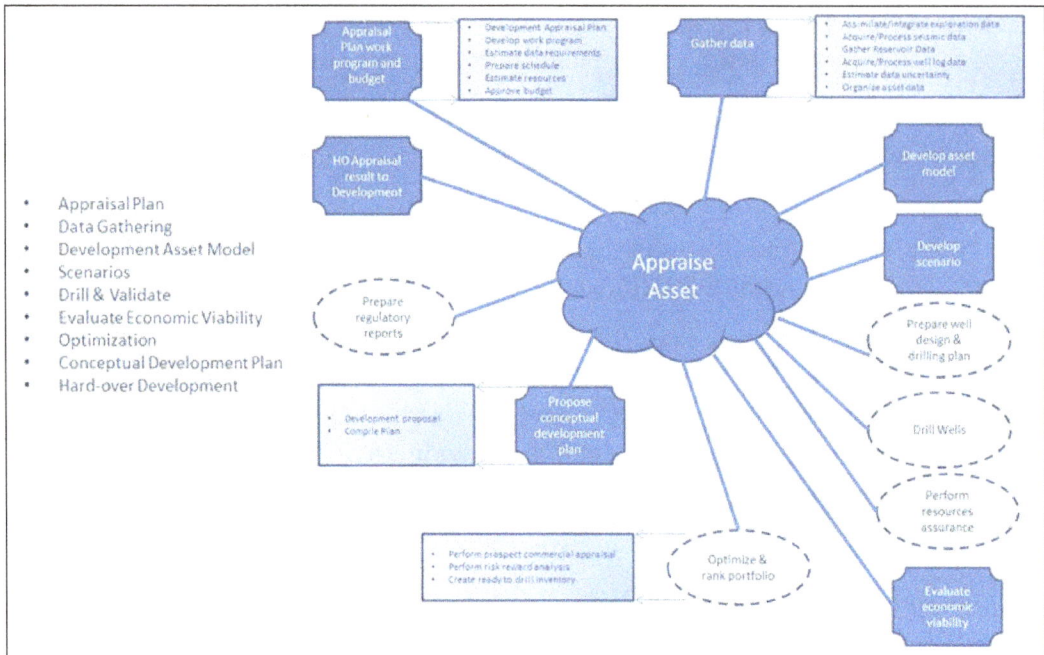

Goals of the appraisal phase.

Reservoir Engineering is a branch of petroleum engineering that applies scientific principles to the exploitation of oil and gas reservoir to obtain a high economic recovery. Reservoir Engineering analyzes the production potential of the reservoir and determine the technical ways and means that should be used to optimize oil or gas recovery.

Reservoir engineers makes description of the reservoir from the available data and refine these data by applying the laws of physics to forecast reservoir behavior during production and depletion.

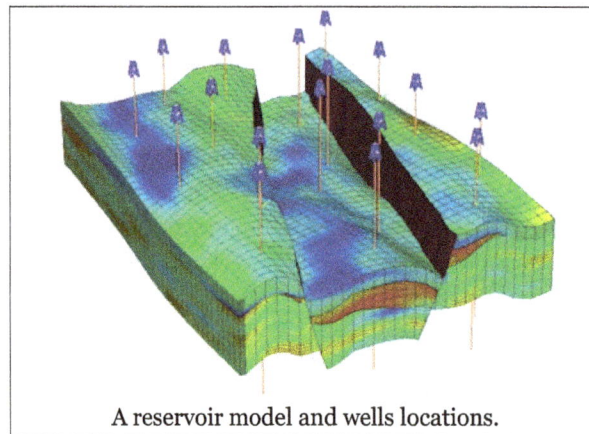

A reservoir model and wells locations.

Reservoir engineers work out development scenarios along with precise recommendations for the number and positioning of the wells, the drilling schedule, the production profile, etc.

Activities of the appraisal phase include:

- Planning and execution of a data acquisition program (seismic).

- Reprocessing existing seismic data.

- Drilling of appraisal wells.

- Evaluation of the results obtained from the seismic and drilling activities.

- Use of the data update reservoir models.

- Carry out initial development planning and an environmental impact assessment (EIA) study.

Field Development Phase

The development stage takes place after successfully completing the appraisal period and before the beginning of the field production. Field Development Plans (FDPs) provide the necessary support for field optimization, and include all activities and processes required to optimally develop a field.

In general, development activities and processes involve:

- Environmental impact, geophysics, geology, reservoir and production engineering, infrastructure, well design and construction, completion design, surface facilities, economics and risk assessment.

In particular, the activities and people involved in the development stage:

- Define a precise Field Development Plan (FDP) – geologists, geophysicists and reservoir engineers.

- Decide the best production/injection well placement and design – drilling engineers, reservoir engineers, geologists.

- Select the optimal production facilities required to properly process hydrocarbons before their treatment – production engineers, reservoir engineers, facilities engineers.

- Choose the transport options and route to export oil and gas – logistics engineers.

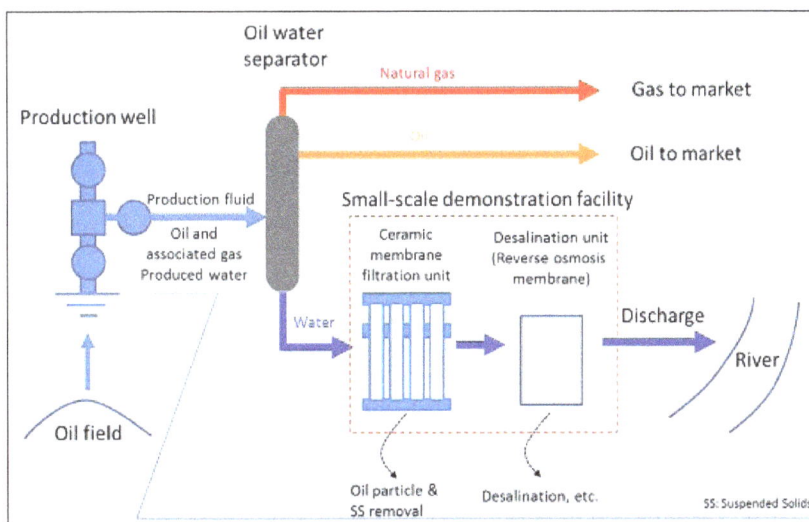

Typical Process Plant for Field Development – Technip.

The development of an oil and gas field costs millions of dollars and may require long time (5-10 years) to be fully realized.

Costs and duration of the development phase depends on the location of the field, the size and complexity of the facilities, and the number of wells needed to achieve the production and economic targets.

A FDP must consider:

- Objectives of the development.
- Petroleum engineering data.
- Operating and maintenance objectives.
- Description of the engineering facilities.
- Cost and manpower estimates.
- Project planning budget proposal.

Once the field development plan is approved and before the beginning of production, some actions follow:

- Detailed design of the facilities.
- Procurement of the materials of construction.
- Fabrication of the facilities.
- Installation of the facilities.
- Commissioning of all plant and equipment.

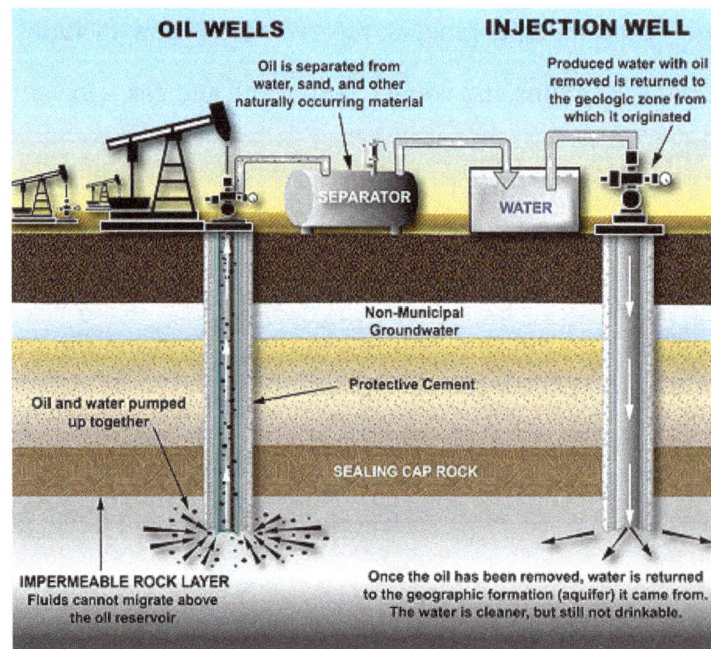

An example of an offshore field development.

Development Planning

After obtaining mining concession rights and confirming the existence of crude oil, a development plan is prepared. Based on the evaluation of reservoir analysis results, the development planning of the field including the facilities planning is established to optimize the crude production. In this planning, crude recovery is maximized considering the production profile, crude properties change over the production lifetime Additional development may be required in the interim of production phase. It is important to optimize the development cost over the exploration and production life, including initial costs required until the start of production, the development period prior to production and the facility expansion, to accommodate the change of the production profile during the life of production.

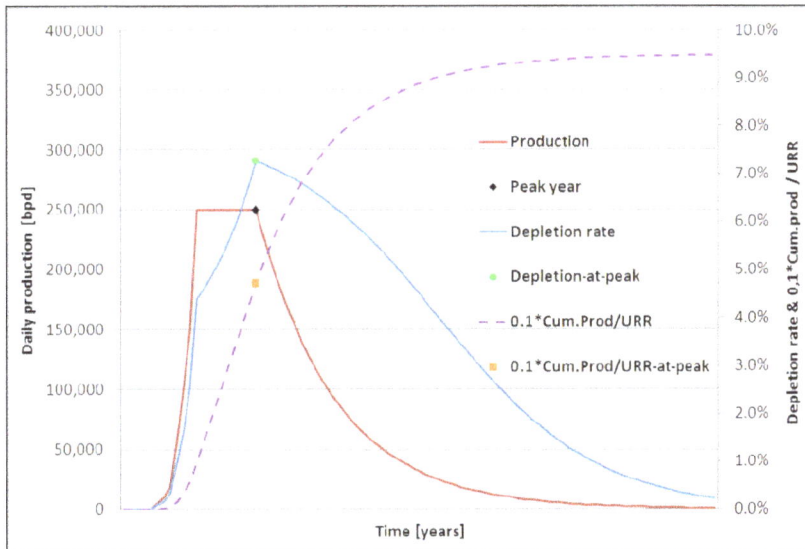

Sample for basic data of oil field development.

Oil Field Development Planning and Implementation

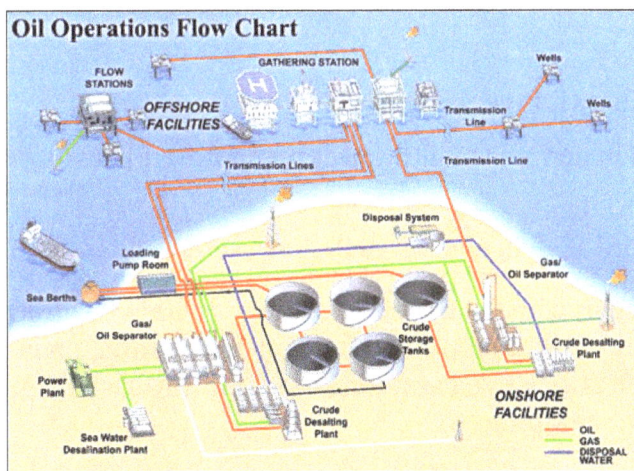

Oil production facility.

The production profile of the oil field is clearly identified by drilling exploration wells, and conceptual planning geared toward production is performed while revising the development plan prepared in the initial stages. It is important that the plan in the initial stages has sufficient tolerance, because the development conditions and other factors have not yet been fully understood. However, too much wider tolerance could increase the initial costs required. At the same time, it is also important to study the environmental impact in the greater interests of global environmental conservation. Since the reservoir performance differs according to each oil field, experience and successful outcomes related to oil well development are necessary when establishing a development planning. Furthermore, detailed consideration is necessary not only for the change in crude oil production and composition over the production lifetime but also for seasonal and daily fluctuation.

Production Monitoring and Secondary and Tertiary Recovery

After production starts, production monitoring comes to play an important role. It is important to monitor and record the moisture content of the product, the properties of the crude oil and the pressure decline during the production. Such reservoir performance monitored provide the basis for the selection criterion for appropriate secondary and tertiary recovery methods.

In general, the production process moves from primary production, secondary recovery, and to tertiary recovery. Primary production means using natural energy stored in the reservoir as a driving mechanism for production. Secondary recovery would imply adding some energy to the reservoir by injecting fluids such as water or gas, to help support the reservoir pressure as production take place. Tertiary recovery, known as Enhanced Oil Recovery (EOR), seek for production techniques after primary and secondary recovery methods. Production and recovery methods shall be optimized based on the reservoir dynamic behavior over the lifetime of the production.

Petroleum Production Phase

Production phase aims to the recovery of the reservoir fluids to surface followed by their processing.

All production and maintenance activities are carried out to meet strict safety and environmental policies and procedures.

During the production phase it is necessary to:

* Control production and injection to meet up approved plans for volumes and quality of products.
* Monitor and record all data to manage the reservoir, wells and facilities.

The production phase begins with the first commercial volumes of hydrocarbons ("first oil") flowing through the wellhead – this establishes a fundamental turning point from a cash flow stand point. From now on, cash is generated and can be used to pay back the previous investments, or may be made available for new projects.

The development and production planning are based on the expected production profile which are strongly dependent on the reservoir drive mechanism. It is the forecasted production profile that will determine the size of facilities required to treat and disposes the fluids and the number and phasing of wells to be drilled.

A typical production profile is made up by three phases:

- Build-up period:
  - During this period production wells are progressively brought on flow.

- Plateau period:
  - A constant production rate is mantained.

- Decline period:
  - All producers show declining of production rates.

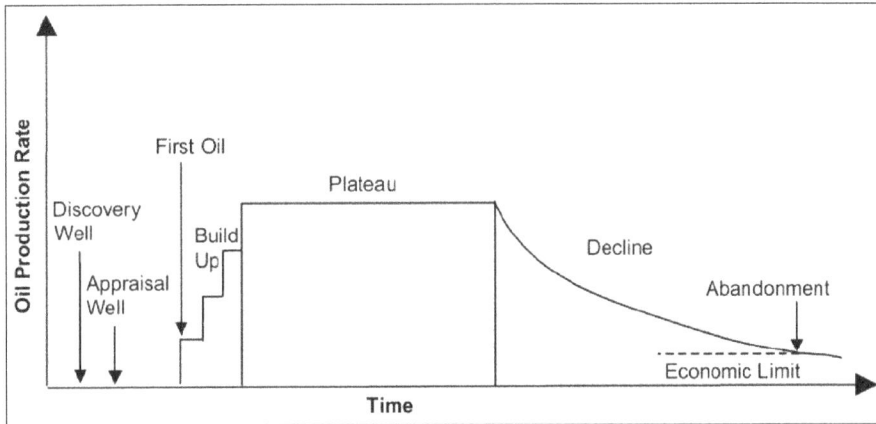

Typical field production profile.

The mode of operations and maintenance required are specified in the field development plan.

It is necessary to deliver product at the required rate and quality and therefore, the product quality specification and contract terms guide the activities of the production operations.

The "product" quality is not limited to oil and gas, but also to other fluid/byproduct streams.

The process and facilities engineers design:

- The equipment for a range of capacities (maximum throughputs).
- The type of operation and maintenance.

The processing plants are designed to treat specific types of hydrocarbon according to their characteristics. The recovered fluids are processed for technical, safety and economic reasons.

By processing the produced fluids, it is possible to:

- Remove any fluid non-hydrocarbon components (water, carbon dioxide, hydrogen sulphide, nitrogen, etc.) Or solids.
- Bring the hydrocarbons quality up to the required specifications.

Gas processing may include:

- Separation.

- Dehydration for the removal of water carried out through various systems such as refrigeration, adsorption and absorption.

- Condensate removal carried out by means of refrigeration and adsorption.

- Desulphurisation-decarbonation to remove hydrogen sulphide and carbon dioxide (when found in intolerable quantities).

Crude oil processing may include:

- Gas-oil separation.

- Water-oil separation.

- Desalting to eliminate the saline content.

- Desulphurization to remove the hydrogen sulphide content.

- Stabilization to meet the required standards.

Typical crude oil processing.

During production, monitoring and control of the production processes is performed by a combination of instrumentation and control equipment which:

- Increase production rates.

- Reduce operating expenditure.

- Reduce capital expenditure.

- Increase safety.

- Improve environmental protection.

Waste disposal is another important aspect of the production process and it should cover all effluent streams. The treatment of these products and waste is discussed jointly with the Environment department, process and facilities engineers. An oil and gas processing plant include the utilities systems which do not take part to the processes but support production operations.

Utility systems include:

- Power system (fuel gas and diesel).

- Water and potable water treatment system.

- Chemicals and lubrication oils alarm.

- Shut-down system.

- Fire protection and fire fighting system.

- Instrument/utility air system, etc.

The necessary control and monitoring of processing operations is managed from the control room, which is a dedicated structure where operators perform plant operations monitor and control using sophisticated and automatic control systems in a safe and efficient manner.

Operators in a processing plant control room.

Components of the Petroleum Production System

Volume and Phase of Reservoir Hydrocarbons

The reservoir consists of one or several interconnected geological flow units. While the shape of a well and converging flow have created in the past the notion of radial flow configuration, modern techniques such as 3-D seismic and new logging and well testing measurements allow for a more precise description of the shape of a geological flow unit and the ensuing production character of the well. This is particularly true in identifying lateral and vertical boundaries and the inherent heterogeneities.

Appropriate reservoir description, including the extent of heterogeneities, discontinuities, and anisotropies, while always important, has become compelling after the emergence of horizontal wells and complex well architecture with total lengths of reservoir exposure of many thousands of feet.

Figure is a schematic showing two wells, one vertical and the other horizontal, contained within a reservoir with potential lateral heterogeneities or discontinuities (sealing faults), vertical boundaries (shale lenses), and anisotropies (stress or permeability).

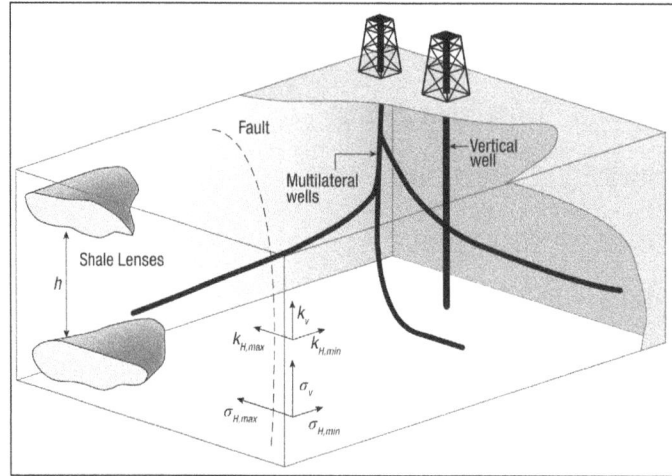

Common reservoir heterogeneities, anisotropies, discontinuities, and boundaries affecting the performance of vertical, horizontal, and complex-architecture wells.

While appropriate reservoir description and identification of boundaries, heterogeneities, and anisotropies is important, it is somewhat forgiving in the presence of only vertical wells. These issues become critical when horizontal and complex wells are drilled.

The encountering of lateral discontinuities (including heterogeneous pressure depletion caused by existing wells) has a major impact on the expected complex well production. The well branch trajectories vis à vis the azimuth of directional properties also has a great effect on well production. Ordinarily, there would be only one set of optimum directions.

Understanding the geological history that preceded the present hydrocarbon accumulation is essential. There is little doubt that the best petroleum engineers are those who understand the geological processes of deposition, fluid migration, and accumulation. Whether a reservoir is an anticline, a fault block, or a channel sand not only dictates the amount of hydrocarbon present but also greatly controls well performance.

Porosity

All of petroleum engineering deals with the exploitation of fluids residing within porous media. Porosity, simply defined as the ratio of the pore volume, V_p, to the bulk volume, V_b, is an indicator of the amount of fluid in place. Porosity values vary from over 0.3 to less than 0.1.

$$\varphi = \frac{V_p}{V_b}$$

The porosity of the reservoir can be measured based on laboratory techniques using reservoir cores or with field measurements including logs and well tests. Porosity is one of the very first measurements obtained in any exploration scheme, and a desirable value is essential for the continuation of any further activities toward the potential exploitation of a reservoir. In the absence of substantial porosity there is no need to proceed with an attempt to exploit a reservoir.

Reservoir Height

Often known as "reservoir thickness" or "pay thickness," the reservoir height describes the thickness of a porous medium in hydraulic communication contained between two layers. These layers are usually considered impermeable. At times the thickness of the hydrocarbon-bearing formation is distinguished from an underlaying water-bearing formation, or aquifer. Often the term "gross height" is employed in a multilayered, but co-mingled during production, formation. In such cases the term "net height" may be used to account for only the permeable layers in a geologic sequence.

Well logging techniques have been developed to identify likely reservoirs and quantify their vertical extent. For example, measuring the spontaneous potential (SP) and knowing that sandstones have a distinctly different response than shales (a likely lithology to contain a layer), one can estimate the thickness of a formation. Figure is a well log showing clearly the deflection of the spontaneous potential of a sandstone reservoir and the clearly different response of the adjoining shale layers. This deflection corresponds to the thickness of a potentially hydrocarbon-bearing, porous medium.

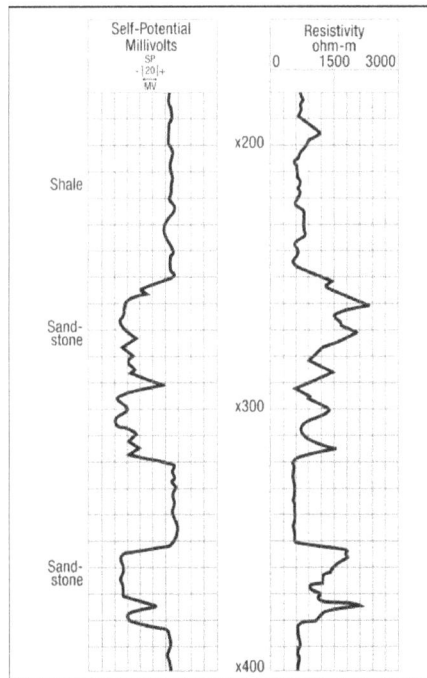

Spontaneous potential and electrical resistivity logs
identifying sandstones versus shales, and water-bearing
versus hydrocarbon-bearing formations.

The presence of satisfactory net reservoir height is an additional imperative in any exploration activity.

Fluid Saturations

Oil and/or gas are never alone in "saturating" the available pore space. Water is always present. Certain rocks are "oil-wet" implying that oil molecules cling to the rock surface. More frequently, rocks are "water-wet." Electrostatic forces and surface tension act to create these wettabilities, which may change, usually with detrimental consequences, as a result of injection of fluids, drilling, stimulation, or other activity, and in the presence of surface-acting chemicals. If the water is present but does not flow, the corresponding water saturation is known as "connate" or "interstitial." Saturations larger than this value would result in free flow of water along with hydrocarbons.

Petroleum hydrocarbons, which are mixtures of many compounds, are divided into oil and gas. Any mixture depending on its composition and the conditions of pressure and temperature may appear as liquid (oil) or gas or a mixture of the two.

Frequently the use of the terms oil and gas is blurred. Produced oil and gas refer to those parts of the total mixture that would be in liquid and gaseous states, respectively, after surface separation. Usually the corresponding pressure and temperature are "standard conditions," that is, usually (but not always) 14.7 psi and 60° F.

Flowing oil and gas in the reservoir imply, of course, that either the initial reservoir pressure or the induced flowing bottomhole pressures are such as to allow the concurrent presence of two phases. Temperature, except in the case of high-rate gas wells, is for all practical purposes constant.

An attractive hydrocarbon saturation is the third critical variable (along with porosity and reservoir height) to be determined before a well is tested or completed. A classic method, currently performed in a variety of ways, is the measurement of the formation electrical resistivity. Knowing that formation brines are good conductors of electricity (i.e., they have poor resistivity) and hydrocarbons are the opposite, a measurement of this electrical property in a porous formation of sufficient height can detect the presence of hydrocarbons. With proper calibration, not just the presence but also the hydrocarbon saturation (i.e., fraction of the pore space occupied by hydrocarbons) can be estimated.

Figure above also contains a resistivity log. The SP log along with the resistivity log, showing a high resistivity within the same zone, are good indicators that the identified porous medium is likely saturated with hydrocarbons.

The combination of porosity, reservoir net thickness, and saturations is essential in deciding whether a prospect is attractive or not. These variables can allow the estimation of hydrocarbons near the well.

Classification of Reservoirs

All hydrocarbon mixtures can be described by a phase diagram such as the one shown in Figure. Plotted are temperature (x axis) and pressure (y axis). A specific point is the *critical point*, where the properties of liquid and gas converge. For each temperature less than the critical-point temperature (to the left of T_c in Figure) there exists a pressure called the "bubble-point" pressure, above which only liquid (oil) is present and below which gas and liquid coexist. For lower pressures (at constant temperature), more gas is liberated. Reservoirs above the bubble-point pressure are called "undersaturated."

If the initial reservoir pressure is less than or equal to the bubble-point pressure, or if the flowing bottomhole pressure is allowed to be at such a value (even if the initial reservoir pressure is above the bubble point), then free gas will at least form and will likely flow in the reservoir. This type of a reservoir is known as "two-phase" or "saturated."

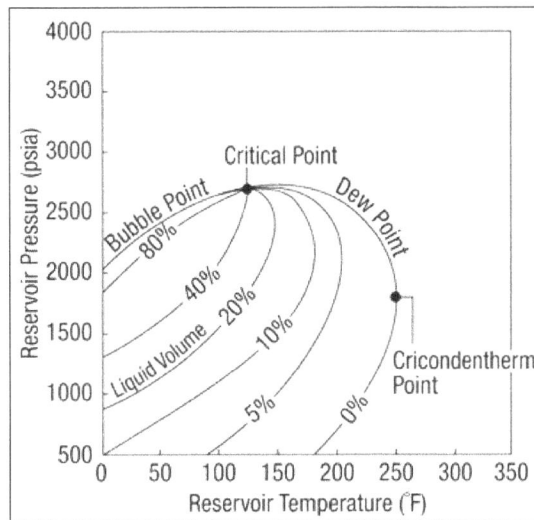

Oilfield hydrocarbon phase diagram showing bubble-point and
dew-point curves, lines of constant-phase distribution, region of
retrograde condensation, and the critical and cricondentherm points.

For temperatures larger than the critical point (to the right of T_c in figure), the curve enclosing the two-phase envelop is known as the "dew-point" curve. Outside, the fluid is gas, and reservoirs with these conditions are "lean" gas reservoirs.

The maximum temperature of a two-phase envelop is known as the "cricondentherm." Between these two points there exists a region where, because of the shape of the gas saturation curves, as the pressure decreases, liquid or "condensate" is formed. This happens until a limited value of the pressure, after which further pressure reduction results in revaporization. The region in which this phenomenon takes place is known as the "retrograde condensation" region, and reservoirs with this type of behavior are known as "retrograde condensate reservoirs."

Each hydrocarbon reservoir has a characteristic phase diagram and resulting physical and thermodynamic properties. These are usually measured in the laboratory with tests performed on fluid samples obtained from the well in a highly specialized manner. Petroleum thermodynamic properties are known collectively as PVT (pressure–volume–temperature) properties.

Areal Extent

Favorable conclusions on the porosity, reservoir height, fluid saturations, and pressure (and implied phase distribution) of a petroleum reservoir, based on single well measurements, are insufficient for both the decision to develop the reservoir and for the establishment of an appropriate exploitation scheme.

Advances in 3-D and wellbore seismic techniques, in combination with well testing, can increase greatly the region where knowledge of the reservoir extent (with height, porosity, and saturations) is possible. Discontinuities and their locations can be detected. As more wells are drilled, additional information can enhance further the knowledge of the reservoir's peculiarities and limits.

The areal extent is essential in the estimation of the "original-oil (or gas)-in-place." The hydrocarbon volume, V_{HC}, in reservoir cubic ft is,

$$V_{HC} = Ah\phi(1-S_w)$$

where, A is the areal extent in ft2, h is the reservoir thickness in ft, ϕ is the porosity, and Sw is the water saturation. (Thus, 1 − Sw is the hydrocarbon saturation.) The porosity, height, and saturation can of course vary within the areal extent of the reservoir.

Equation ($V_{HC} = Ah\phi(1-S_w)$) can lead to the estimation of the oil or gas volume under standard conditions after dividing by the oil formation volume factor, B_o, or the gas formation volume factor, B_g. This factor is simply a ratio of the volume of liquid or gas under reservoir conditions to the corresponding volumes under standard conditions. Thus, for oil,

$$N = \frac{7758\,Ah\phi(1-S_w)}{B_o}$$

where, N is in stock tank barrels (STB). In Equation ($N = \frac{7758\,Ah\phi(1-S_w)}{B_o}$) the area is in acres. For gas,

$$G = \frac{Ah\phi(1-S_w)}{B_g}$$

where, G is in standard cubic ft (SCF) and A is in ft².

The gas formation volume factor (traditionally, res ft³/SCF), B_g, simply implies a volumetric relationship and can be calculated readily with an application of the real gas law. The gas formation volume factor is much smaller than 1.

The oil formation volume factor (res bbl/STB), B_o, is not a simple physical property. Instead, it is an empirical thermodynamic relationship allowing for the reintroduction into the liquid (at the elevated reservoir pressure) of all of the gas that would be liberated at standard conditions. Thus the oil formation volume factor is invariably larger than 1, reflecting the swelling of the oil volume because of the gas dissolution.

Permeability

The presence of a substantial porosity usually (but not always) implies that pores will be interconnected. Therefore the porous medium is also "permeable." The property that describes the ability of fluids to flow in the porous medium is permeability. In certain lithologies (e.g., sandstones), a larger porosity is associated with a larger permeability. In other lithologies (e.g., chalks), very large porosities, at times over 0.4, are not necessarily associated with proportionately large permeabilities.

Correlations of porosity versus permeability should be used with a considerable degree of caution, especially when going from one lithology to another. For production engineering calculations these correlations are rarely useful, except when considering matrix stimulation. In this instance, correlations of the altered permeability with the altered porosity after stimulation are useful.

The concept of permeability was introduced by Darcy (1856) in a classic experimental work from which both petroleum engineering and groundwater hydrology have benefited greatly.

Figure is a schematic of Darcy's experiment. The flow rate (or fluid velocity) can be measured against pressure (head) for different porous media.

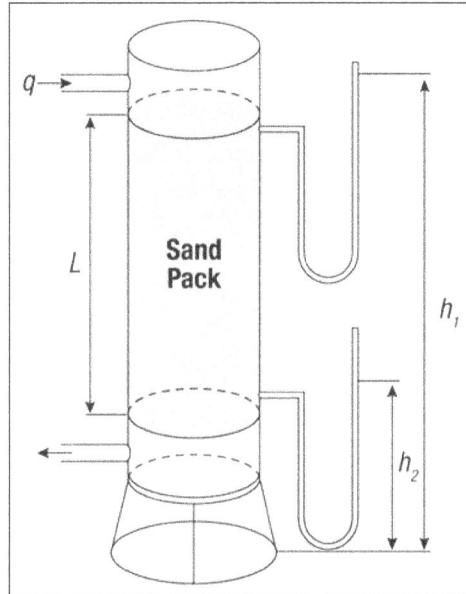

Darcy's experiment. Water flows through a sand pack
and the pressure difference (head) is recorded.

Darcy observed that the flow rate (or velocity) of a fluid through a specific porous medium is linearly proportional to the head or pressure difference between the inlet and the outlet and a characteristic property of the medium. Thus,

$$u \, \alpha \, k \, \Delta \, p$$

Where k is the permeability and is a characteristic property of the porous medium. Darcy's experiments were done with water. If fluids of other viscosities flow, the permeability must be divided by the viscosity and the ratio k/μ is known as the "mobility."

The Zone Near the Well, the Sandface and the Well Completion

The zone surrounding a well is important. First, even without any man-made disturbance, converging, radial flow results in a considerable pressure drop around the wellbore and, as will be demonstrated the pressure drop away from the well varies logarithmically with the distance. This means that the pressure drop in the first foot away from the well is naturally equal to that 10 feet away and equal to that 100 feet away, and so on. Second, all intrusive activities such as drilling, cementing, and well completion are certain to alter the condition of the reservoir near the well. This is usually detrimental and it is not inconceivable that in some cases 90% of the total pressure drop in the reservoir may be consumed in a zone just a few feet away from the well.

Matrix stimulation is intended to recover or even improve the near-wellbore permeability. (There is damage associated even with stimulation. It is the net effect that is expected to be beneficial.)

Hydraulic fracturing, today one of the most widely practiced well-completion techniques, alters the manner by which fluids flow to the well; one of the most profound effects is that near-well radial flow and the damage associated with it are eliminated.

Many wells are cemented and cased. One of the purposes of cementing is to support the casing, but at formation depths the most important reason is to provide zonal isolation. Contamination of the produced fluid from the other formations or the loss of fluid *into* other formations can be envisioned readily in an open-hole completion. If no zonal isolation or wellbore stability problems are present, the well can be open hole. A cemented and cased well must be perforated in order to reestablish communication with the reservoir. Slotted liners can be used if a cemented and cased well is not deemed necessary and are particularly common in horizontal wells where cementing is more difficult.

Finally, to combat the problems of sand or other fines production, screens can be placed between the well and the formation. Gravel packing can be used as an additional safeguard and as a means to keep permeability-reducing fines away from the well.

The various well completions and the resulting near-wellbore zones are shown in figure.

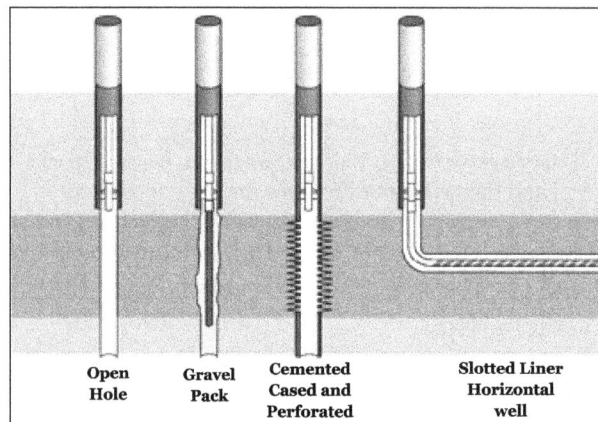

Options for well completions.

The ability to direct the drilling of a well allows the creation of highly deviated, horizontal, and complex wells. In these cases, a longer to far longer exposure of the well with the reservoir is accomplished than would be the case for vertical wells.

The Well

Entrance of fluids into the well, following their flow through the porous medium, the near-well zone, and the completion assembly, requires that they are lifted through the well up to the surface.

There is a required flowing pressure gradient between the bottomhole and the well head. The pressure gradient consists of the potential energy difference (hydrostatic pressure) and the frictional pressure drop. The former depends on the reservoir depth and the latter depends on the well length.

If the bottomhole pressure is sufficient to lift the fluids to the top, then the well is "naturally flowing." Otherwise, artificial lift is indicated. Mechanical lift can be supplied by a pump. Another technique is to reduce the density of the fluid in the well and thus to reduce the hydrostatic pressure. This is accomplished by the injection of lean gas in a designated spot along the well. This is known as "gas lift."

The Surface Equipment

After the fluid reaches the top, it is likely to be directed toward a manifold connecting a number of wells. The reservoir fluid consists of oil, gas (even if the flowing bottomhole pressure is larger than the bubble-point pressure, gas is likely to come out of solution along the well), and water.

Traditionally, the oil, gas, and water are not transported long distances as a mixed stream, but instead are separated at a surface processing facility located in close proximity to the wells. An exception that is becoming more common is in some offshore fields, where production from subsea wells, or sometimes the commingled production from several wells, may be transported long distances before any phase separation takes place.

Finally, the separated fluids are transported or stored. In the case of formation water it is usually disposed in the ground through a reinjection well.

The reservoir, well, and surface facilities are sketched in figure.

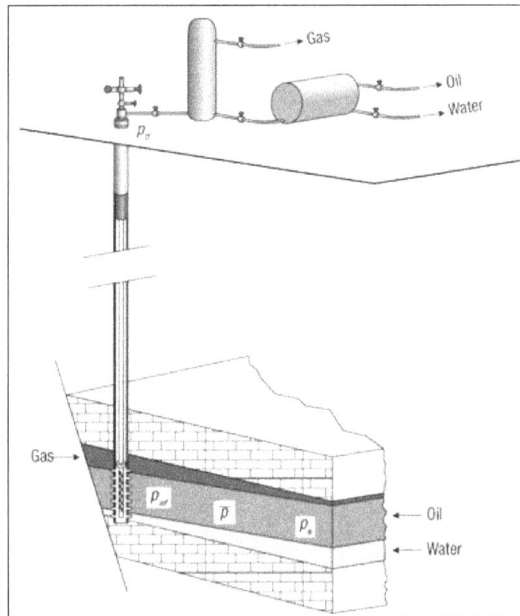

The petroleum production system, including the reservoir, underground well completion, the well, wellhead assembly, and surface facilities.

Well Abandonment Phase

The end of the life of the field occurs when it is no longer economic for the operator to continue production, and consequently the wells will be plugged and abandoned. Well workovers or stimulation, artificial lift, drilling new wells, etc, may help in prolonging the field's life, but they are undertaken if they can be economically justified by increased production.

Well Plug and Abandonment (P and A)

The basic of P and A operations vary little, whether the well is on land or offshore, and the oil and gas industry has developed methods and materials designed to provide long term zonal isolation even when downhole change over the time.

The objective of all P and A operations is to achieve the following:

- Isolate and protect all fresh and near fresh water zones.

- Isolate and protect all future commercial zones.

- Prevent leaks in perpetuity from or into the well.

- Remove surface equipment and cut pipe to a mandated level below the surface.

A traditional abandonment process begins with a well killing operation in which produced fluids are circulated out of the well, or bull headed into the formation, and replaced by drilling fluids heavy enough to contain any open formation pressures. Xmas tree is removed and replaced by a blowout preventer, through which the production tubing can be removed. Cement is then pumped and placed across the open perforations and squeezed into the formation to seal off all productive layers.

Depending on the well configuration, a series of cement and wireline plugs in both the liner and production casing will be set to a depth level with the top of cement behind the production casing.

The production casing is cut and removed above the top of cement, and a cement plug positioned over the casing stub to isolate the annulus and any formation which may still be open below the intermediate casing shoe.

Offshore Facilities Decommissioning

Different types of offshore production facilities have different options for decommissioning:

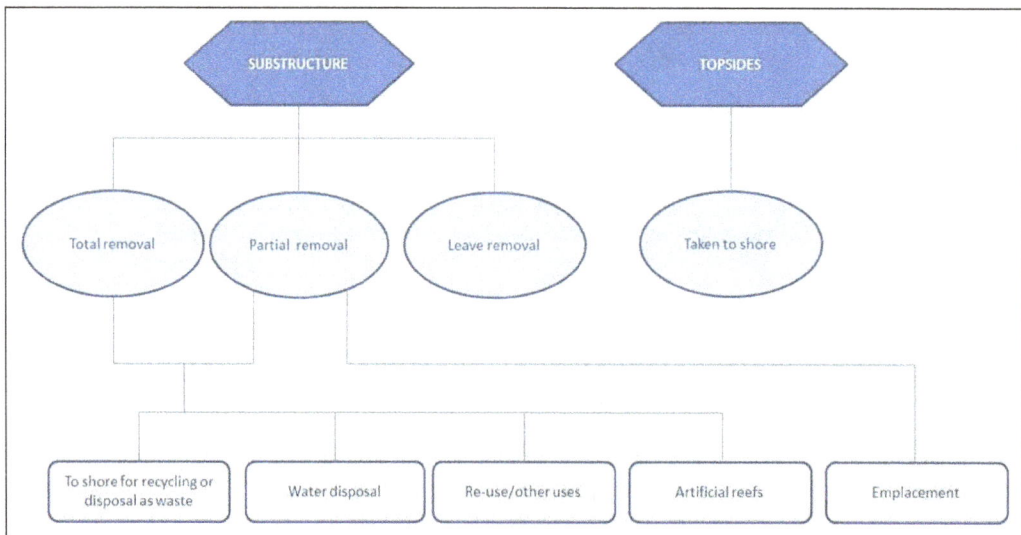

Decommissioning options.

Decommissioning works.

Onshore Facilities Decommissioning

Onshore processing facilities have to be cleaned of all hazardous compounds and scrapped.

The land under the facilities may also have to be reconditioned, and environmental restoration at the original conditions must be carried out.

In some cases, before abandoning a hydrocarbon reservoir (gas field), it is worth examining the possibility of transforming it into a gas storage reservoir. But not all depleted gas fields are suited for this "transformation". Moreover, the properties of the reservoir must be such that the gas injected for storage can be produced without losses, and the reservoirs must be able to guarantee productivity.

References

- Heshelow, Kathy (2010). Investing in Oil and Gas: The ABC's of DPPs. Iuniverse. p. 52. ISBN 978-1450261715

- Petroleum-exploration, upstream: oil-gasportal.com, Retrieved 28 june, 2019

- Smil, Vaclav (2017). Oil - A Beginner's Guide. Oneworld Publications. ISBN 978-1786072863

- "Ensign Launches Newest And Most Powerful Automated ADR 1500S Pad Drill Rigs In Montney Play", New Tech Magazine, Calgary, Alberta, 21 November 2014, archived from the original on 10 December 2014, retrieved 6 December 2014

- Krah, Jaclyn; Unger, Richard L. (7 August 2013). "The Importance of Occupational Safety and Health: Making for a "Super" Workplace". National Institute for Occupational Safety and Health. Retrieved 16 January 2015

Chapter 5

Petroleum Refinery

The industrial process plant where the transformation and refinement of crude oil into gasoline, naphtha, diesel fuel, kerosene, jet fuel, liquefied petroleum gas, etc. takes place is known as a petroleum refinery. All the diverse processes which take place in a petroleum refinery have been carefully analyzed in this chapter.

The petroleum refining industry converts crude oil into more than 2500 refined products, including liquefied petroleum gas, gasoline, kerosene, aviation fuel, diesel fuel, fuel oils, lubricating oils, and feedstocks for the petrochemical industry. Petroleum refinery activities start with receipt of crude for storage at the refinery, include all petroleum handling and refining operations, and they terminate with storage preparatory to shipping the refined products from the refinery.

The petroleum refining industry employs a wide variety of processes. A refinery's processing flow scheme is largely determined by the composition of the crude oil feedstock and the chosen slate of petroleum products. The example refinery flow scheme presented in figure shows the general processing arrangement used by refineries in the United States for major refinery processes. The arrangement of these processes will vary among refineries, and few, if any, employ all of these processes.

Petroleum Refinery Process

Petroleum refining begins with the distillation, or fractionation, of crude oils into separate hydrocarbon groups. The resultant products are directly related to the characteristics of the crude oil being processed. Most of these products of distillation are further converted into more useable products by changing their physical and molecular structures through cracking, reforming and other conversion processes. These products are subsequently subjected to various treatment and separation processes, such as extraction, hydrotreating and sweetening, in order to produce finished products. Whereas the simplest refineries are usually limited to atmospheric and vacuum distillation, integrated refineries incorporate fractionation, conversion, treatment and blending with lubricant, heavy fuels and asphalt manufacturing; they may also include petrochemical processing.

The first refinery, which opened in 1861, produced kerosene by simple atmospheric distillation. Its by-products included tar and naphtha. It was soon discovered that high-quality lubricating oils could be produced by distilling petroleum under vacuum. However, for the next 30 years, kerosene was the product consumers wanted most. The two most significant events which changed this situation were:

- The invention of the electric light, which decreased the demand for kerosene.

- The invention of the internal-combustion engine, which created a demand for diesel fuel and gasoline (naphtha).

With the advent of mass production and the First World War, the number of gasoline-powered vehicles increased dramatically, and the demand for gasoline grew accordingly. However, only a certain amount of gasoline could be obtained from crude oil through atmospheric and vacuum distillation processes. The first thermal cracking process was developed in 1913. Thermal cracking subjected heavy fuels to both pressure and intense heat, physically breaking their large molecules into smaller ones, producing additional gasoline and distillate fuels. A sophisticated form of thermal cracking, visbreaking, was developed in the late 1930s to produce more desirable and valuable products.

As higher-compression gasoline engines were developed, there was a demand for higher-octane gasoline with better anti-knock characteristics. The introduction of catalytic cracking and polymerization processes in the mid- to late 1930s met this demand by providing improved gasoline yields and higher octane numbers. Alkylation, another catalytic process, was developed in the early 1940s to produce more high-octane aviation gasoline and petrochemical feedstocks, the starting materials, for explosives and synthetic rubber. Subsequently, catalytic isomerization was developed to convert hydrocarbons to produce increased quantities of alkylation feedstocks.

Following the Second World War, various reforming processes were introduced which improved gasoline quality and yield, and produced higher-quality products. Some of these involved the use of catalysts and/or hydrogen to change molecules and remove sulphur. Improved catalysts, and process methods such as hydrocracking and reforming, were developed throughout the 1960s to increase gasoline yields and improve anti-knock characteristics. These catalytic processes also produced molecules with a double bond (alkenes), forming the basis of the modern petrochemical industry.

The numbers and types of different processes used in modern refineries depend primarily on the nature of the crude feedstock and finished product requirements. Processes are also affected by economic factors including crude costs, product values, availability of utilities and transportation. The chronology of the introduction of various processes is given in table below.

Table: Summary of the history of refining processing.

Year	Process name	Process purpose	Process by-products
1862	Atmospheric distillation	Produce kerosene	Naphtha, tar, etc.
1870	Vacuum distillation	Lubricants (original) Cracking feedstocks (1930s)	Asphalt, residual Coker feedstocks
1913	Thermal cracking	Increase gasoline	Residual, bunker fuel
1916	Sweetening	Reduce sulphur and odour	Sulphur
1930	Thermal reforming	Improve octane number	Residual
1932	Hydrogenation	Remove sulphur	Sulphur
1932	Coking	Produce gasoline base stocks	Coke
1933	Solvent extraction	Improve lubricant viscosity index	Aromatics
1935	Solvent dewaxing	Improve pour point	Waxes
1935	Catalytic polymerization	Improve gasoline yield and octane number	Petrochemical feedstocks
1937	Catalytic cracking	Higher octane gasoline	Petrochemical feedstocks

1939	Visbreaking	Reduce viscosity	Increased distillate, tar
1940	Alkylation	Increase gasoline octane and yield	High-octane aviation gasoline
1940	Isomerization	Produce alkylation feedstock	Naphtha
1942	Fluid catalytic cracking	Increase gasoline yield and octane	Petrochemical feedstocks
1950	Deasphalting	Increase cracking feedstock	Asphalt
1952	Catalytic reforming	Convert low-quality naphtha	Aromatics
1954	Hydrodesulphurization	Remove sulphur	Sulphur
1956	Inhibitor sweetening	Remove mercaptan	Disulphides
1957	Catalytic isomerization	Convert to molecules with high octane number	Alkylation feedstocks
1960	Hydrocracking	Improve quality and reduce sulphur	Alkylation feedstocks
1974	Catalytic dewaxing	Improve pour point	Wax
1975	Residual hydrocracking	Increase gasoline yield from residual	Heavy residuals

Basic Refining Processes and Operations

Petroleum refining processes and operations can be classified into the following basic areas: separation, conversion, treatment, formulating and blending, auxiliary refining operations and refining non-process operations. See figure below for a simplified flow chart.

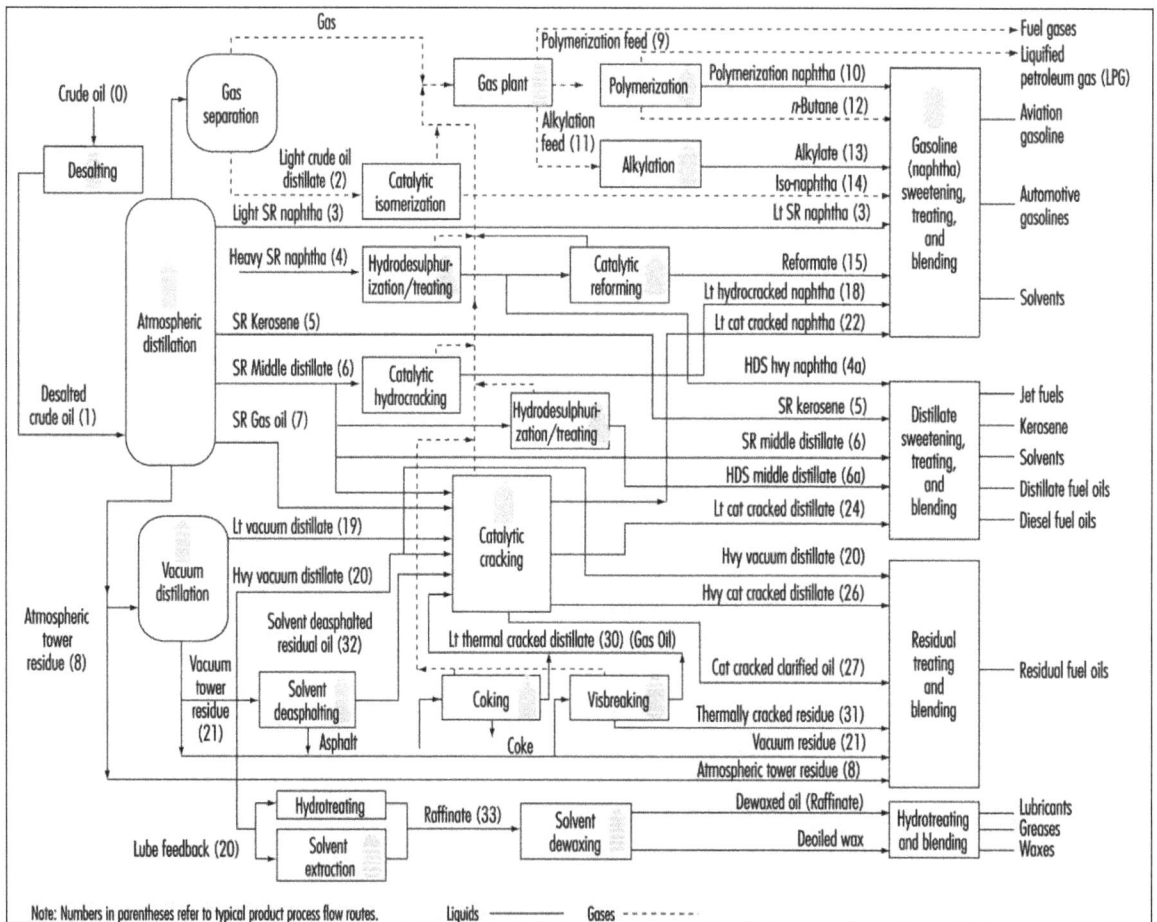

Refinery process chart.

Separation: Crude oil is physically separated by fractionation in atmospheric and vacuum distillation towers, into groups of hydrocarbon molecules with various boiling-point ranges, called "fractions" or "cuts".

Conversion: Conversion processes used to change the size and/or structure of hydrocarbon molecules include:

- Decomposition (dividing) by hydro-thermal and catalytic cracking, coking and visbreaking.

- Unification (combining) through alkylation and polymerization.

- Alteration (rearranging) with isomerization and catalytic reforming.

- Treatment.

Since the beginning of refining, various treatment methods have been used to remove non-hydrocarbons, impurities and other constituents that adversely affect the performance properties of finished products or reduce the efficiency of the conversion processes. Treatment involves both chemical reactions and physical separation, such as dissolving, absorption or precipitation, using a variety and combination of processes. Treatment methods include removing or separating aromatics and naphthenes, as well as removing impurities and undesirable contaminants. Sweetening compounds and acids are used to desulphurize crude oil before processing, and to treat products during and after processing. Other treatment methods include crude desalting, chemical sweetening, acid treating, clay contacting, hydrodesulphurizing, solvent refining, caustic washing, hydrotreating, drying, solvent extraction and solvent dewaxing.

Formulating and blending is the process of mixing and combining hydrocarbon fractions, additives and other components to produce finished products with specific desired performance properties.

Auxiliary refining operations: Other refinery operations which are required to support hydrocarbon processing include light ends recovery; sour water stripping; solid waste, waste water and process water treatment and cooling; hydrogen production; sulphur recovery; and acid and tail gas treatment. Other process functions are providing catalysts, reagents, steam, air, nitrogen, oxygen, hydrogen and fuel gases.

Refinery non-process facilities: All refineries have a multitude of facilities, functions, equipment and systems which support the hydrocarbon process operations. Typical support operations are heat and power generation; product movement; tank storage; shipping and handling; flares and relief systems; furnaces and heaters; alarms and sensors; and sampling, testing and inspecting. Non-process facilities and systems include firefighting, water and protection systems, noise and pollution controls, laboratories, control rooms, warehouses, maintenance and administrative facilities.

Major Products of Crude Oil Refining

Petroleum refining has evolved continuously in response to changing consumer demand for better and different products. The original process requirement was to produce kerosene as a cheaper and better source of fuel for lighting than whale oil. The development of the internal combustion engine led to the production of benzene, gasoline and diesel fuels. The evolution of the airplane created a need for high-octane aviation gasoline and jet fuel, which is a sophisticated form of the original refinery product, kerosene. Present-day refineries produce a variety of products, including many which are used as feedstocks for cracking processes and lubricant manufacturing, and for the petrochemical industry. These products can be broadly classified as fuels, petrochemical feedstocks, solvents, process oils, lubricants and special products such as wax, asphalt and coke.

Table: Principal products of crude oil refining.

Hydrocarbon gases	Uses
Liquified gases	Cooking and industrial gas Motor fuel gas Illuminating gas Ammonia Synthetic fertilizer Alcohols Solvents and acetone Plasticizers Resins and fibres for plastics and textiles Paints and varnish
Chemical industry feedstock	Rubber products
Carbon black	Printing inks Rubber industry
Light distillates	
Light naphthas	Olefins Solvents and diluents Extraction solvents Chemical industry feedstocks
Intermediate naphthas	Aviation and motor gasoline Dry-cleaning solvents
Heavy naphthas	Military jet fuel Jet fuel and kerosene Tractor fuel
Gas oil	Cracking stock Heating oil and diesel fuel Metallurgical fuel Absorber oil—benzene and gasoline recovery
Heavy distillates	
Technical oils	Textile oils Medicinal oils and cosmetics White oil—food industry
Lubricating oils	Transformer and spindle oils Motor and engine oils Machine and compressor oils Turbine and hydraulic oils Transmission oils Equipment and cable insulation oils Axle, gear and steam engine oils Metal treating, cutting and grinding oils Quenching and rust inhibitor oils Heat transfer oils Lubricating greases and compounds Printing ink oils

Paraffin wax	Rubber industry Pharmaceuticals and cosmetics Food and paper industries Candles and matches
Residues	
Petrolatum	Petroleum jelly Cosmetics Rust inhibitors and lubricants Cable coating compounds
Residual fuel oil	No. 6 boiler and process fuel oil
Asphalts	Paving asphalt Roofing materials Asphaltic lubricants Insulating and foundation protection Waterproof paper products
Refinery by-products	
Coke	Electrodes and fuel
Sulphonates	Emulsifiers
Sulphuric acid	Synthetic fertilizer
Sulphur	Chemicals
Hydrogen	Hydrocarbon reformation

A number of chemicals are used in, or formed as a result of, hydrocarbon processing. A brief description of those which are specific and pertinent to refining follows:

Sulphur Dioxide

Flue gas from burning high-sulphur-content fuels usually contains high levels of sulphur dioxide, which usually is removed by water scrubbing.

Caustics

Caustics are added to desalting water to neutralize acids and reduce corrosion. Caustics are also added to desalted crude in order to reduce the amount of corrosive chlorides in the tower overheads. They are used in refinery treating processes to remove contaminants from hydrocarbon streams.

Nitrogen Oxides and Carbon Monoxide

Flue gas contains up to 200 ppm of nitric oxide, which reacts slowly with oxygen to form nitrogen dioxide. Nitric oxide is not removed by water scrubbing, and nitrogen dioxide can dissolve in water to form nitrous and nitric acid. Flue gas normally contains only a slight amount of carbon monoxide, unless combustion is abnormal.

Hydrogen Sulphide

Hydrogen sulphide is found naturally in most crude oils and is also formed during processing by the decomposition of unstable sulphur compounds. Hydrogen sulphide is an extremely toxic, colourless, flammable gas which is heavier than air and soluble in water. It has a rotten egg odour which is discernible at concentrations well below its very low exposure limit. This smell cannot be relied upon to provide adequate warning as the senses are almost immediately desensitized upon exposure. Special detectors are required to alert workers to the presence of hydrogen sulphide, and proper respiratory protection should be used in the presence of the gas. Exposure to low levels of hydrogen sulphide will cause irritation, dizziness and headaches, while exposure to levels in excess of the prescribed limits will cause nervous system depression and eventually death.

Sour Water

Sour water is process water which contains hydrogen sulphide, ammonia, phenols, hydrocarbons and low-molecular-weight sulphur compounds. Sour water is produced by steam stripping hydro-carbon fractions during distillation, regenerating catalyst, or steam stripping hydrogen sulphide during hydrotreating and hydrofinishing. Sour water is also generated by the addition of water to processes to absorb hydrogen sulphide and ammonia.

Sulphuric Acid and Hydrofluoric Acid

Sulphuric acid and hydrofluoric acid are used as catalysts in alkylation processes. Sulphuric acid is also used in some of the treatment processes.

Solid Catalysts

A number of different solid catalysts in many forms and shapes, from pellets to granular beads to dusts, made of various materials and having various compositions, are used in refining processes. Extruded pellet catalysts are used in moving and fixed bed units, while fluid bed processes use fine, spherical particulate catalysts. Catalysts used in processes which remove sulphur are impregnated with cobalt, nickel or molybdenum. Cracking units use acid-function catalysts, such as natural clay, silica alumina and synthetic zeolites. Acid-function catalysts impregnated with platinum or other noble metals are used in isomerization and reforming. Used catalysts require special han-dling and protection from exposures, as they may contain metals, aromatic oils, carcinogenic poly-cyclic aromatic compounds or other hazardous materials, and may also be pyrophoric.

Fuels

The principal fuel products are liquefied petroleum gas, gasoline, kerosene, jet fuel, diesel fuel and heating oil and residual fuel oils.

Liquefied petroleum gas (LPG), which consists of mixtures of paraffinic and olefinic hydrocarbons such as propane and butane, is produced for use as a fuel, and is stored and handled as liquids under pressure. LPG has boiling points ranging from about −74 °C to +38 °C, is colourless, and the vapours are heavier than air and extremely flammable. The important qualities from an occu-pational health and safety perspective of LPGs are vapour pressure and control of contaminants.

Gasoline: The most important refinery product is motor gasoline, a blend of relatively low-boiling hydrocarbon fractions, including reformate, alkylate, aliphatic naphtha (light straight-run naph-tha), aromatic naphtha (thermal and catalytic cracked naphtha) and additives. Gasoline blending stocks have boiling points which range from ambient temperatures to about 204 °C, and a flash-point below −40 °C. The critical qualities for gasoline are octane number (anti-knock), volatility (starting and vapour lock) and vapour pressure (environmental control). Additives are used to enhance gasoline performance and provide protection against oxidation and rust formation. Aviation gasoline is a high-octane product, specially blended to perform well at high altitudes.

Tetra ethyl lead (TEL) and tetra methyl lead (TML) are gasoline additives which improve octane ratings and anti-knock performance. In an effort to reduce lead in automotive exhaust emissions, these additives are no longer in common use, except in aviation gasoline.

Ethyl tertiary butyl ether (ETBE), methyl tertiary butyl ether (MTBE), tertiary amyl methyl ether (TAME) and other oxygenated compounds are used in lieu of TEL and TML to improve unleaded gasoline anti-knock performance and reduce carbon monoxide emissions.

Jet fuel and kerosene: Kerosene is a mixture of paraffins and naphthenes with usually less than 20% aromatics. It has a flashpoint above 38 °C and a boiling range of 160 °C to 288 °C, and is used for lighting, heating, solvents and blending into diesel fuel. Jet fuel is a middle distillate kerosene product whose critical qualities are freezepoint, flashpoint and smokepoint. Commercial jet fuel has a boiling range of about 191 °C to 274 °C, and military jet fuel from 55 °C to 288 °C.

Distillate fuels: Diesel fuels and domestic heating oils are light-coloured mixtures of paraffins, naphthenes and aromatics, and may contain moderate quantities of olefins. Distillate fuels have flashpoints above 60 °C and boiling ranges of about 163 °C to 371 °C, and are often hydrodesulphurized for improved stability. Distillate fuels are combustible and when heated may emit vapours which can form ignitable mixtures with air. The desirable qualities required for distillate fuels include controlled flash- and pourpoints, clean burning, no deposit formation in storage tanks, and a proper diesel fuel cetane rating for good starting and combustion.

Residual fuels: Many ships and commercial and industrial facilities use residual fuels or combinations of residual and distillate fuels, for power, heat and processing. Residual fuels are dark-coloured, highly viscous liquid mixtures of large hydrocarbon molecules, with flashpoints above 121 °C and high boiling points. The critical specifications for residual fuels are viscosity and low sulphur content (for environmental control).

Health and Safety Considerations

The primary safety hazard of LPG and gasoline is fire. The high volatility and high flammability of the lower-boiling-point products allows vapours to evaporate readily into air and form flammable mixtures which can be easily ignited. This is a recognized hazard that requires specific storage, containment and handling precautions, and safety measures to assure that releases of vapours and sources of ignition are controlled so that fires do not occur. The less volatile fuels, such as kerosene and diesel fuel, should be handled carefully to prevent spills and possible ignition, as their vapours are also combustible when mixed with air in the flammable range. When working in atmospheres containing fuel vapours, concentrations of highly volatile, flammable product vapours in air are often restricted to no more than 10% of the lower flammable limits (LFL), and concentrations of less volatile, combustible product vapours to no more than 20% LFL, depending on applicable company and government regulations, in order to reduce the risk of ignition.

Although gasoline vapour levels in air mixtures are typically maintained below 10% of the LFL for safety purposes, this concentration is considerably above the exposure limits to be observed for health reasons. When inhaled, small amounts of gasoline vapour in air, well below the lower flammable limit, can cause irritation, headaches and dizziness, while inhalation of larger concentrations can cause loss of consciousness and eventually death. Long-term health effects may also be possible. Gasoline contains benzene, for example, a known carcinogen with allowable exposure limits of only a few parts per million. Therefore, even working in gasoline vapour atmospheres at levels below 10% LFL requires appropriate industrial hygiene precautions, such as respiratory protection or local exhaust ventilation.

In the past, many gasolines contained tetra-ethyl or tetra methyl alky lead anti-knock additives, which are toxic and present serious lead absorption hazards by skin contact or inhalation. Tanks or vessels which contained leaded gasoline at any time during their use must be vented, thoroughly cleaned, tested with a special "lead-in-air" test device and certified to be lead-free to assure that workers can enter without using self-contained or supplied breathing air equipment, even though oxygen levels are normal and the tanks now contain unleaded gasoline or other products.

Gaseous petroleum fractions and the more highly volatile fuel products have a mild anaesthetic effect, generally in inverse ratio to molecular weight. Lower-boiling-point liquid fuels, such as gasoline and kerosene, produce a severe chemical pneumonitis if inhaled, and should not be siphoned by mouth or accidentally ingested. Gases and vapours may also be present in sufficiently high concentrations to displace oxygen (in the air) below normal breathing levels. Maintaining vapour concentrations below the exposure limits and oxygen levels at normal breathing ranges, is usually accomplished by purging or ventilation.

Cracked distillates contain small amounts of carcinogenic polycyclic aromatic hydrocarbons (PAHs); therefore, exposure should be limited. Dermatitis may also develop from exposure to gasoline, kerosene and distillate fuels, as they have a tendency to defat the skin. Prevention is accomplished by use of personal protective equipment, barrier creams or reduced contact and good hygienic practices, such as washing with warm water and soap instead of cleaning hands with gasoline, kerosene or solvents. Some persons have skin sensitivity to the dyes used to colour gasoline and other distillate products.

Residual fuel oils contain traces of metals and may have entrained hydrogen sulphide, which is extremely toxic. Residual fuels which have high cracked stocks boiling above 370 °C contain carcinogenic PAHs. Repeated exposure to residual fuels without appropriate personal protection, should be avoided, especially when opening tanks and vessels, as hydrogen sulphide gas may be emitted.

Petrochemical Feedstocks

Many products derived from crude-oil refining, such as ethylene, propylene and butadiene, are olefinic hydrocarbons derived from refinery cracking processes, and are intended for use in the petrochemical industry as feedstocks for the production of plastics, ammonia, synthetic rubber, glycol and so on.

Petroleum Solvents

A variety of pure compounds, including benzene, toluene, xylene, hexane and heptane, whose boiling points and hydrocarbon composition are closely controlled, are produced for use as solvents.

Solvents may be classified as aromatic or non-aromatic, depending on their composition. Their use as paint thinners, dry-cleaning fluids, degreasers, industrial and pesticide solvents and so on, is generally determined by their flashpoints, which vary from well below −18 °C to above 60 °C.

The hazards associated with solvents are similar to those of fuels in that the lower flashpoint solvents are flammable and their vapours, when mixed with air in the flammable range, are ignitable. Aromatic solvents will usually have more toxicity than non-aromatic solvents.

Process Oils

Process oils include the high boiling range, straight run atmospheric or vacuum distillate streams and those which are produced by catalytic or thermal cracking. These complex mixtures, which contain large paraffinic, naphthenic and aromatic hydrocarbon molecules with more than 15 carbon atoms, are used as feedstocks for cracking or lubricant manufacturing. Process oils have fairly high viscosities, boiling points ranging from 260 °C to 538 °C, and flashpoints above 121 °C.

Process oils are irritating to the skin and contain high concentrations of PAHs as well as sulphur, nitrogen and oxygen compounds. Inhalation of vapours and mists should be avoided, and skin exposure should be controlled by the use of personal protection and good hygienic practices.

Lubricants and Greases

Lubricating oil base stocks are produced by special refining processes to meet specific consumer requirements. Lubricating base stocks are light- to medium-coloured, low-volatile, medium- to high-viscous mixtures of paraffinic, naphthenic and aromatic oils, with boiling ranges from 371 °C to 538 °C. Additives, such as demulsifiers, anti-oxidants and viscosity improvers, are blended into the lubricating oil base stocks to provide the characteristics required for motor oils, turbine and hydraulic oils, industrial greases, lubricants, gear oils and cutting oils. The most critical quality for lubricating oil base stock is a high viscosity index, providing for less change in viscosity under varying temperatures. This characteristic may be present in the crude oil feed stock or attained through the use of viscosity index improver additives. Detergents are added to keep in suspension any sludge formed during the use of the oil.

Greases are mixtures of lubricating oils and metallic soaps, with the addition of special-purpose materials such as asbestos, graphite, molybdenum, silicones and talc to provide insulation or lubricity. Cutting and metal-process oils are lubricating oils with special additives such as chlorine, sulphur and fatty-acid additives which react under heat to provide lubrication and protection to the cutting tools. Emulsifiers and bacteria prevention agents are added to water-soluble cutting oils.

Although lubricating oils by themselves are non-irritating and have little toxicity, hazards may be presented by the additives. Users should consult supplier material safety data information to determine the hazards of specific additives, lubricants, cutting oils and greases. The primary lubricant hazard is dermatitis, which can usually be controlled by the use of personal protective equipment together with proper hygienic practices. Occasionally workers may develop a sensitivity to cutting oils or lubricants which will require reassignment to a job where contact cannot occur. There are some concerns about carcinogenic exposure to mists from naphthenic-based cutting and light spindle oils, which can be controlled by substitution, engineering controls or personal protection. The hazards of exposure to grease are similar to those of lubricating oil, with the addition of any hazards presented by the grease materials or additives.

Special Products

- Wax is used for protecting food products; in coatings; as an ingredient in other products such as cosmetics and shoe polish and for candles.

- Sulphur is produced as a result of petroleum refining. It is stored either as a heated, molten liquid in closed tanks or as a solid in containers or outdoors.

- Coke is almost pure carbon, with a variety of uses from electrodes to charcoal briquettes, depending on its physical characteristics, which result from the coking process.

- Asphalt, which is primarily used for paving roads and roofing materials, should be inert to most chemicals and weather conditions.

Waxes and asphalts are solid at ambient temperatures, and higher temperatures are needed for storage, handling and transportation, with the resulting hazard of burns. Petroleum wax is so highly refined that it usually does not present any hazards. Skin contact with wax can lead to plugging of pores, which can be controlled by proper hygienic practices. Exposure to hydrogen sulphide when asphalt and molten sulphur tanks are opened can be controlled by the use of appropriate engineering controls or respiratory protection. Sulphur is also readily ignitable at elevated temperatures.

Petroleum Refining Processes

Hydrocarbon refining is the use of chemicals, catalysts, heat and pressure to separate and combine the basic types of hydrocarbon molecules naturally found in crude oil into groups of similar molecules. The refining process also rearranges the structures and bonding patterns of the basic molecules into different, more desirable hydrocarbon molecules and compounds. The type of hydrocarbon (paraffinic, naphthenic or aromatic) rather than the specific chemical compounds present, is the most significant factor in the refining process.

Throughout the refinery, operations procedures, safe work practices and the use of appropriate personal protective clothing and equipment, including approved respiratory protection, is needed for fire, chemical, particulate, heat and noise exposures and during process operations, sampling, inspection, turnaround and maintenance activities. As most refinery processes are continuous and the process streams are contained in enclosed vessels and piping, there is limited potential for exposure. However, the potential for fire exists because even though refinery operations are closed processes, if a leak or release of hydrocarbon liquid, vapour or gas occurs, the heaters, furnaces and heat exchangers throughout the process units are sources of ignition.

Crude Oil Pretreatment

Desalting

Crude oil often contains water, inorganic salts, suspended solids and water-soluble trace metals.

The first step in the refining process is to remove these contaminants by desalting (dehydration) in order to reduce corrosion, plugging and fouling of equipment, and to prevent poisoning the catalysts in processing units. Chemical desalting, electrostatic separation and filtering are three typical methods of crude-oil desalting. In chemical desalting, water and chemical surfactants (demulsifiers) are added to the crude oil, heated so that salts and other impurities dissolve into the water or attach to the water, and are then held in a tank where they settle out. Electrical desalting applies high-voltage electrostatic charges in order to concentrate suspended water globules in the bottom portion of the settling tank. Surfactants are added only when the crude oil has a large amount of suspended solids. A third, less common process involves filtering heated crude oil using diatomaceous earth as a filtration medium.

In chemical and electrostatic desalting, the crude feedstock is heated to between 66 °C and 177 °C, to reduce viscosity and surface tension for easier mixing and separation of the water. The temperature is limited by the vapour pressure of the crude-oil feedstock. Both methods of desalting are continuous. Caustic or acid may be added to adjust the pH of the water wash, and ammonia added to reduce corrosion. Waste water, together with contaminants, is discharged from the bottom of the settling tank to the waste water treatment facility. The desalted crude oil is continuously drawn from the top of the settling tanks and sent to an atmospheric crude distillation (fractionating) tower.

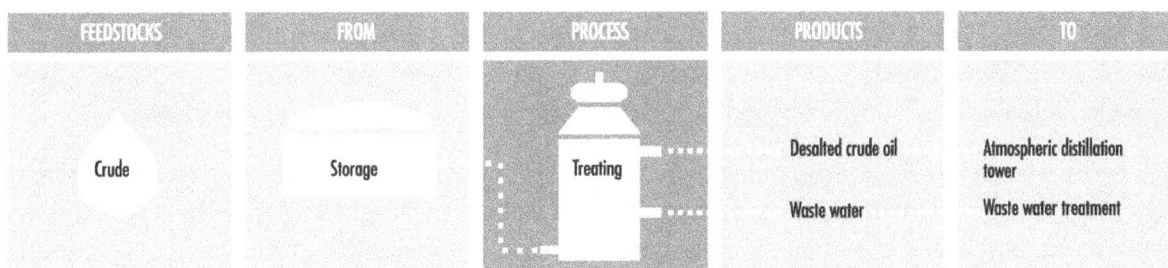

Desalting (pre-treatment) process.

Inadequate desalting causes fouling of heater tubes and heat exchangers in all refinery process units, restricting product flow and heat transfer, and resulting in failures due to increased pressures and temperatures. Overpressuring the desalting unit will cause failure.

Corrosion, which occurs due to the presence of hydrogen sulphide, hydrogen chloride, naphthenic (organic) acids and other contaminants in the crude oil, also causes equipment failure. Corrosion occurs when neutralized salts (ammonium chlorides and sulphides) are moistened by condensed water. Because desalting is a closed process, there is little potential for exposure to crude oil or process chemicals, unless a leak or release occurs. A fire may occur as a result of a leak in the heaters, allowing a release of low-boiling-point components of crude oil.

There is the possibility of exposure to ammonia, dry chemical demulsifiers, caustics and/or acids during desalting. Where elevated operating temperatures are used when desalting sour crude oils, hydrogen sulphide will be present. Depending on the crude feedstock and the treatment chemicals used, the waste water will contain varying amounts of chlorides, sulphides, bicarbonates, ammonia, hydrocarbons, phenol and suspended solids. If diatomaceous earth is used in filtration, exposures should be minimized or controlled since diatomaceous earth can contain silica with a very fine particle size, making it a potential respiratory hazard.

Crude Oil Separation Processes

The first step in petroleum refining is the fractionation of crude oil in atmospheric and vacuum distillation towers. Heated crude oil is physically separated into various fractions, or straight-run cuts, differentiated by specific boiling-point ranges and classified, in order of decreasing volatility, as gases, light distillates, middle distillates, gas oils and residuum. Fractionation works because the gradation in temperature from the bottom to the top of the distillation tower causes the higher-boiling-point components to condense first, while the lower-boiling-point fractions rise higher in the tower before they condense. Within the tower, the rising vapours and the descending liquids (reflux) mix at levels where they have compositions in equilibrium with each other. Special trays are located at these levels (or stages) which remove a fraction of the liquid which condenses at each level. In a typical two-stage crude unit, the atmospheric tower, producing light fractions and distillate, is immediately followed by a vacuum tower which processes the atmospheric residuals. After distillation, only a few hydrocarbons are suitable for use as finished products without further processing.

Atmospheric Distillation

In atmospheric distillation towers, the desalted crude feedstock is preheated using recovered process heat. It then flows to a direct-fired crude charge heater, where it is fed into the vertical distillation column just above the bottom at pressures slightly above atmosphere and at temperatures from 343 °C to 371 °C, to avoid undesirable thermal cracking at higher temperatures. The lighter (lower boiling point) fractions diffuse into the upper part of the tower, and are continuously drawn off and directed to other units for further processing, treating, blending and distribution.

Fractions with the lowest boiling points, such as fuel gas and light naphtha, are removed from the top of the tower by an overhead line as vapours. Naphtha, or straight-run gasoline, is taken from the upper section of the tower as an overhead stream. These products are used as petrochemical and reformer feedstocks, gasoline blending stocks, solvents and LPGs.

Intermediate boiling range fractions, including gas oil, heavy naphtha and distillates, are removed from the middle section of the tower as side streams. These are sent to finishing operations for use as kerosene, diesel fuel, fuel oil, jet fuel, catalytic cracker feedstock and blending stocks. Some of these liquid fractions are stripped of their lighter ends, which are returned to the tower as down-flowing reflux streams.

The heavier, higher-boiling-point fractions (called residuum, bottoms or topped crude) which condense or remain at the bottom of the tower, are used for fuel oil, bitumen manufacturing or cracking feedstock, or are directed to a heater and into the vacuum distillation tower for further fractionation.

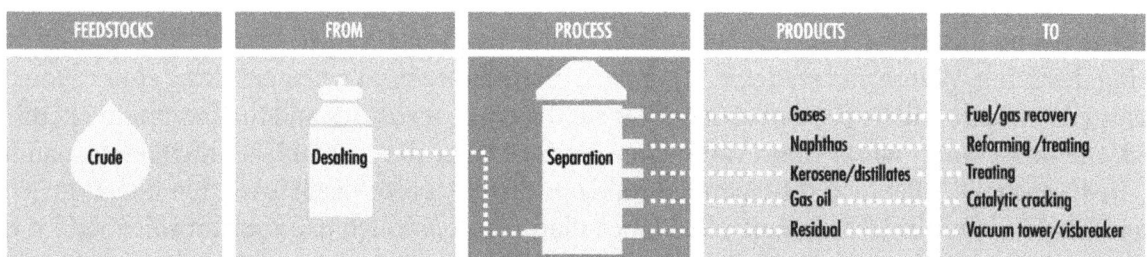

Atmospheric distillation process.

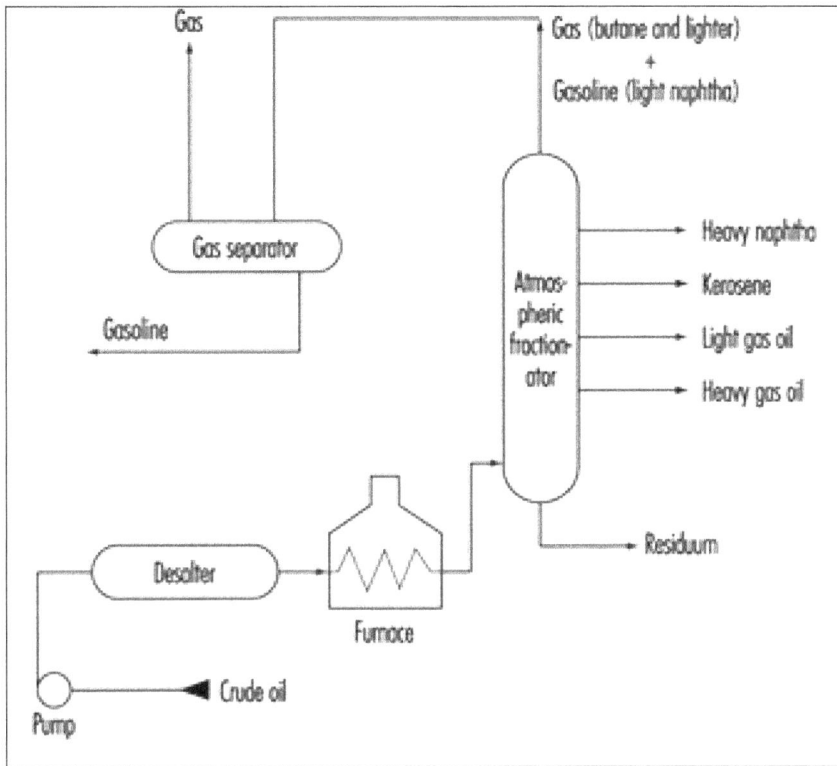

Schematic of atmospheric distrillation process.

Vacuum Distillation

Vacuum distillation towers provide the reduced pressure required to prevent thermal cracking when distilling the residuum, or topped crude, from the atmospheric tower at higher temperatures. The internal designs of some vacuum towers are different from atmospheric towers in that random packing and demister pads are used instead of trays. Larger diameter towers may also be used to keep velocities lower. A typical first-phase vacuum tower may produce gas oils, lubricating oil base stocks and heavy residual for propane deasphalting. A second-phase tower, operating at a lower vacuum, distills surplus residuum from the atmospheric tower which is not used for lube stock processing, and surplus residuum from the first vacuum tower not used for deasphalting.

Vacuum towers are typically used to separate catalytic cracker feedstocks from surplus residuum. Vacuum tower bottoms may also be sent to a coker, used as lubricant or asphalt stock or desulphurized and blended into low-sulphur fuel oil.

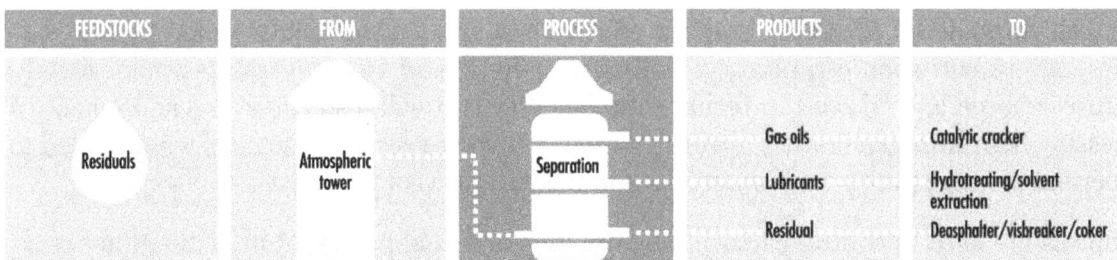

Vacuum distillation process.

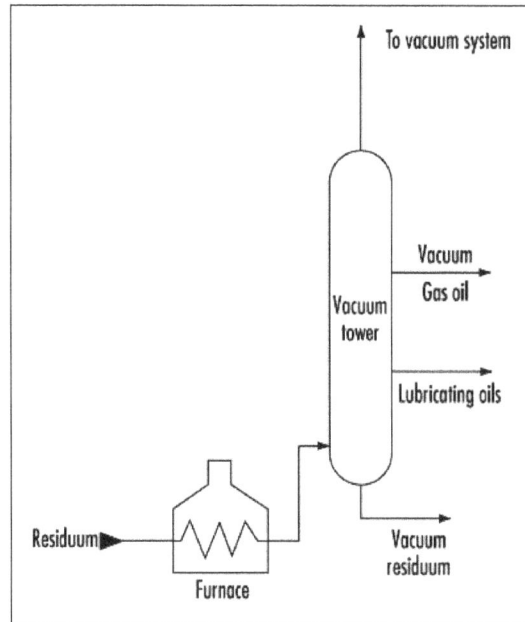

Schematic of vacuum distillation process.

Distillation Columns

Within refineries there are numerous other smaller distillation towers, called columns, designed to separate specific and unique products, which all work on the same principles as atmospheric towers. For example, a depropanizer is a small column designed to separate propane from isobutane and heavier components. Another larger column is used to separate ethyl benzene and xylene. Small "bubbler" towers, called strippers, use steam to remove trace amounts of light products (gasoline) from heavier product streams.

Control temperatures, pressures and reflux must be maintained within operating parameters to prevent thermal cracking from taking place within distillation towers. Relief systems are provided because excursions in pressure, temperature or liquid levels may occur if automatic control devices fail. Operations are monitored in order to prevent crude from entering the reformer charge. Crude feedstocks may contain appreciable amounts of water in suspension which separate during start-up and, along with water remaining in the tower from steam purging, settle in the bottom of the tower. This water may heat to the boiling point and create an instantaneous vaporization explosion upon contact with the oil in the unit.

The preheat exchanger, preheat furnace and bottoms exchanger, atmospheric tower and vacuum furnace, vacuum tower and overhead are susceptible to corrosion from hydrochloric acid (HCl), hydrogen sulphide (H_2S), water, sulphur compounds and organic acids. When processing sour crudes, severe corrosion can occur in both atmospheric and vacuum towers where metal temperatures exceed 232 °C, and in furnace tubing. Wet H_2S will also cause cracks in steel. When processing high-nitrogen crudes, nitrogen oxides, which are corrosive to steel when cooled to low temperatures in the presence of water, form in the flue gases of furnaces.

Chemicals are used to control corrosion by hydrochloric acid produced in distillation units. Ammonia may be injected into the overhead stream prior to initial condensation, and an alkaline solution may be carefully injected into the hot crude oil feed. If sufficient wash water is not injected, deposits of ammonium chloride can form, causing serious corrosion.

Atmospheric and vacuum distillation are closed processes, and exposures are minimal. When sour (high sulphur) crudes are processed, there may be potential exposure to hydrogen sulphide in the preheat exchanger and furnace, tower flash zone and overhead system, vacuum furnace and tower, and bottoms exchanger. Crude oils and distillation products all contain high-boiling aromatic compounds, including carcinogenic PAHs. Short-term exposure to high concentrations of naphtha vapour can result in headaches, nausea and dizziness, and long-term exposure can result in loss of consciousness. Benzene is present in aromatic naphthas, and exposure must be limited. The dehexanizer overhead may contain large amounts of normal hexane, which can affect the nervous system. Hydrogen chloride may be present in the preheat exchanger, tower top zones and overheads. Waste water may contain water-soluble sulphides in high concentrations and other water-soluble compounds, such as ammonia, chlorides, phenol and mercaptan, depending upon the crude feedstock and the treatment chemicals.

Crude Oil Conversion Processes

Conversion processes, such as cracking, combining and rearranging, change the size and structure of hydrocarbon molecules in order to convert fractions into more desirable products.

Table: Overview of petroleum refining processes.

Process name	Action	Method	Purpose	Feedstocks	Products
Fractionation processes					
Atmospheric distillation	Separation	Thermal	Separate fractions	Desalted crude oil	Gas, gas oil, distillate, residual
Vacuum distillation	Separation	Thermal	Separate without cracking	Atmospheric tower residual	Gas oil, lube stock, residual
Conversion processes—Decomposition					
Catalytic cracking	Alteration	Catalytic	Upgrade gasoline	Gas oil, coke distillate	Gasoline, petro-chemical feedstock
Coking	Polymerization	Thermal	Convert vacuum residuals	Residual, heavy oil, tar	Naphtha, gas oil, coke
Hydrocracking	Hydrogenation	Catalytic	Convert to lighter hydrocarbons	Gas oil, cracked oil, residuals	Lighter, higher quality products
Hydrogen steam reforming	Decomposition	Thermal/catalytic	Produce hydro-gen	Desulphurized gas, O_2,steam	Hydrogen, CO,CO_2
Steam cracking	Decomposition	Thermal	Crack large molecules	Atmospheric tower heavy fuel/distillate	Cracked naphtha, coke, residuals
Visbreaking	Decomposition	Thermal	Reduce viscosity	Atmospheric tower residual	Distillate, car
Conversion processes—Unification					
Alkylation	Combining	Catalytic	Unite olefins and isoparaffins	Tower isobutane/cracker olefin	Iso-octane (alkylate)
Grease compounding	Combining	Thermal	Combine soaps and oils	Lube oil, catty acid, alkymetal	Lubricating grease

Polymerization	Polymerization	Catalytic	Unite two or more olefins	Cracker olefins	High octane naphtha, petrochemical stocks
Conversion processes—Alteration/rearrangement					
Catalytic reforming	Alteration/ dehydrogenation	Catalytic	Upgrade low-octane naphtha	Coker/hydrocracker naphtha	High-octane reformate/aromatic
Isomerization	Rearrangement	Catalytic	Convert straight chain to branch	Butane, centane, cexane	Isobutane/pentane/hexane
Treatment processes					
Amine treating	Treatment	Absorption	Remove acidic contaminants	Sour gas, cydrocarbons with CO_2 and H_2S	Acid-free gases and liquid hydrocarbons
Desalting (pre-treatment)	Dehydration	Absorption	Remove contaminants	Crude oil	Desalted crude oil
Drying and sweetening	Treatment	Absorption/ thermal	Remove H_2O and sulphur compounds	Liquid hydrocarbon, LPG, alkylated feedstock	Sweet and dry hydrocarbons
Furfural extraction	Solvent extraction	Absorption	Upgrade middistillate and lubes	Cycle oils and lube feedstocks	High-quality diesel and lube oil
Hydrodesulphurization	Treatment	Catalytic	Remove sulphur, contaminants	High-sulphur residual/gas oil	Desulphurized olefins
Hydrotreating	Hydrogenation	Catalytic	Remove impurities/saturate hydrocarbons	Residuals, cracked hydrocarbons	Cracker feed, cistillate, lube
Phenol extraction	Solvent extraction	Absorption/ thermal	Improve lube viscosity index, colour	Lube oil base stocks	High-quality lube oils
Solvent deasphalting	Treatment	Absorption	Remove asphalt	Vacuum tower residual, cropane	Heavy lube oil, csphalt
Solvent dewaxing	Treatment	Cool/filter	Remove wax from lube stocks	Vacuum tower lube oils	Dewaxed lube base stock
Solvent extraction	Solvent extraction	Absorption/ precipitation	Separate unsaturated aromatics	Gas oil, ceformate, cistillate	High-octane gasoline
Sweetening	Treatment	Catalytic	Remove H_2S, convert mercaptan	Untreated distillate/ gasoline	High-quality distillate/ gasoline

A number of hydrocarbon molecules not normally found in crude oil but important to the refining process are created as a result of conversion. Olefins (alkenes, di-olefins and alkynes) are unsaturated chain- or ring-type hydrocarbon molecules with at least one double bond. They are usually formed by thermal and catalytic cracking and rarely occur naturally in unprocessed crude oil.

Alkenes are straight-chain molecules with the formula CnHn containing at least one double bond (unsaturated) linkage in the chain. The simplest alkene molecule is the mono-olefin ethylene, with two carbon atoms, joined by a double bond, and four hydrogen atoms. Di-olefins (containing two double bonds), such as 1,2-butadiene and 1,3-butadiene, and alkynes (containing a triple bond), such as acetylene, occur in C5 and lighter fractions from cracking. Olefins are more reactive than paraffins or naphthenes, and readily combine with other elements such as hydrogen, chlorine and bromine.

Cracking Processes

Following distillation, subsequent refinery processes are used to alter the molecular structures of the fractions to create more desirable products. One of these processes, cracking, breaks (or cracks) heavier, higher-boiling-point petroleum fractions into more valuable products such as gaseous hydrocarbons, gasoline blending stocks, gas oil and fuel oil. During the process, some of the molecules combine (polymerize) to form larger molecules. The basic types of cracking are thermal cracking, catalytic cracking and hydro-cracking.

Thermal Cracking Processes

Thermal cracking processes, developed in 1913, heat distillate fuels and heavy oils under pressure in large drums until they crack (divide) into smaller molecules with better anti-knock characteristics. This early method, which produced large amounts of solid, unwanted coke, has evolved into modern thermal cracking processes including visbreaking, steam cracking and coking.

Visbreaking

Visbreaking is a mild form of thermal cracking which reduces the pour point of waxy residues and significantly lowers the viscosity of feedstock without affecting its boiling-point range. Residual from the atmospheric distillation tower is mildly cracked in a heater at atmospheric pressure. It is then quenched with cool gas oil to control overcracking, and flashed in a distillation tower. The thermally cracked residue tar, which accumulates in the bottom of the fractionation tower, is vacuum flashed in a stripper and the distillate is recycled.

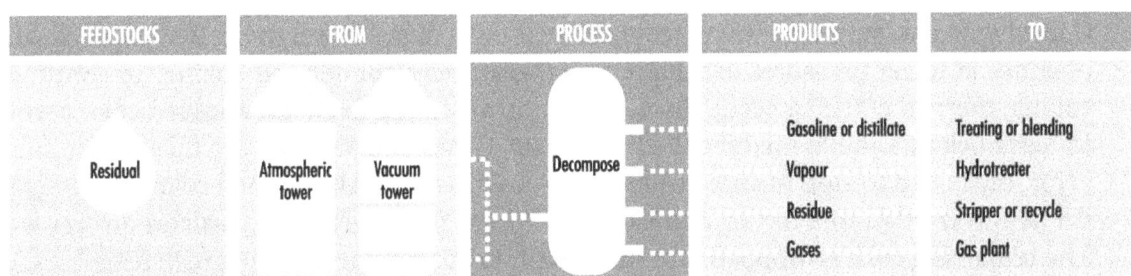

Visbreaking process.

Steam Cracking

Steam cracking produces olefins by thermally cracking large hydrocarbon molecule feedstocks at pressures slightly above atmospheric and at very high temperatures. Residual from steam cracking is blended into heavy fuels. Naphtha produced from steam cracking usually contains benzene, which is extracted prior to hydrotreating.

Coking

Coking is a severe form of thermal cracking used to obtain straight-run gasoline (coker naphtha) and various middle distillate fractions used as catalytic cracking feedstocks. This process so completely reduces hydrogen from the hydrocarbon molecule, that the residue is a form of almost pure carbon called coke. The two most common coking processes are delayed coking and continuous (contact or fluid) coking, which depending upon the reaction mechanism, time, temperature and the crude feedstock, produce three types of coke—sponge, honeycomb and needle coke.

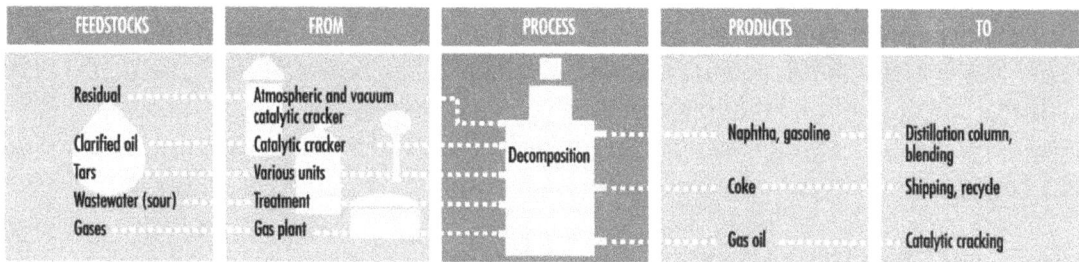

FEEDSTOCKS	FROM	PROCESS	PRODUCTS	TO
Residual	Atmospheric and vacuum catalytic cracker			
Clarified oil	Catalytic cracker		Naphtha, gasoline	Distillation column, blending
Tars	Various units	Decomposition		
Wastewater (sour)	Treatment		Coke	Shipping, recycle
Gases	Gas plant		Gas oil	Catalytic cracking

Coking process.

- Delayed coking: In delayed coking, the feedstock is first charged to a fractionator to separate lighter hydrocarbons, and then combined with heavy recycle oil. The heavy feedstock is fed to the coker furnace and heated to high temperatures at low pressures to prevent premature coking in the heater tubes, producing partial vaporization and mild cracking. The liquid/vapour mixture is pumped from the heater to one or more coker drums, where the hot material is held approximately 24 hours (delayed) at low pressures until it cracks into lighter products. After the coke reaches a predetermined level in one drum, the flow is diverted to another drum to maintain continuous operation. Vapour from the drums is returned to the fractionator to separate out gas, naphtha and gas oils, and to recycle heavier hydrocarbons through the furnace. The full drum is steamed to strip out uncracked hydrocarbons, cooled by water injection and decoked mechanically by an auger rising from the bottom of the drum, or hydraulically by fracturing the coke bed with high-pressure water ejected from a rotating cutter.

- Continuous coking: Continuous (contact or fluid) coking is a moving bed process which operates at lower pressures and higher temperatures than delayed coking. In continuous coking, thermal cracking occurs by using heat transferred from hot recycled coke particles to feedstock in a radial mixer, called a *reactor*. Gases and vapours are taken from the reactor, quenched to stop further reaction and fractionated. The reacted coke enters a surge drum and is lifted to a feeder and classifier where the larger coke particles are removed. The remaining coke is dropped into the reactor preheater for recycling with feedstock. The process is automatic in that there is a continuous flow of coke and feedstock, and coking occurs both in the reactor and in the surge drum.

Health and Safety Considerations

In coking, temperature control should be held within a close range, as high temperatures will produce coke which is too hard to cut out of the drum. Conversely, temperatures which are too low will result in a high asphaltic content slurry. Should coking temperatures get out of control, an exothermic reaction could occur.

In thermal cracking when sour crudes are processed, corrosion can occur where metal temperatures are between 232 °C and 482 °C. It appears that coke forms a protective layer on the metal above 482 °C. However, hydrogen sulphide corrosion occurs when temperatures are not properly controlled above 482 °C. The lower part of the tower, high temperature exchangers, furnace and soaking drums are subject to corrosion. Continuous thermal changes cause coke drum shells to bulge and crack.

Water or steam injection is used to prevent buildup of coke in delayed coker furnace tubes. Water must be completely drained from the coker, so as not to cause an explosion upon recharging with hot coke. In emergencies, alternate means of egress from the working platform on top of coke drums is needed.

Burns may occur when handling hot coke, from steam in the event of a steam line leak, or from hot water, hot coke or hot slurry which may be expelled when opening cokers. The potential exists for exposure to aromatic naphthas containing benzene, hydrogen sulphide and carbon monoxide gases, and to trace amounts of carcinogenic PAHs associated with coking operations. Waste sour water may be highly alkaline, and contain oil, sulphides, ammonia and phenol. When coke is moved as a slurry, oxygen depletion may occur within confined spaces such as storage silos, because wet carbon adsorbs oxygen.

Catalytic Cracking Processes

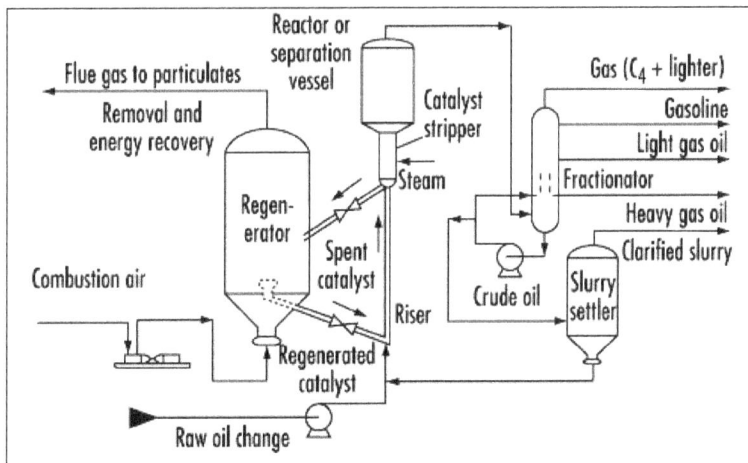

Catalytic cracking breaks up complex hydrocarbons into simpler molecules in order to increase the quality and quantity of lighter, more desirable products and decrease the amount of residuals. Heavy hydrocarbons are exposed at high temperature and low pressure to catalysts which promote chemical reactions. This process rearranges the molecular structure, converting heavy hydrocarbon feedstocks into lighter fractions such as kerosene, gasoline, LPG, heating oil and petrochemical feedstocks. Selection of a catalyst depends upon a combination of the greatest possible reactivity and the best resistance to attrition. The catalysts used in refinery cracking units are typically solid materials (zeolite, aluminium hydrosilicate, treated bentonite clay, Fuller's earth, bauxite and silica-alumina) which are in the form of powders, beads, pellets or shaped materials called extrudites.

Catalytic cracking process.

Schematic of catalytic cracking process.

There are three basic functions in all catalytic cracking processes:

- Reaction—feedstock reacts with catalyst and cracks into different hydrocarbons.

- Regeneration—catalyst is reactivated by burning off coke.

- Fractionation—cracked hydrocarbon stream is separated into various products.

Catalytic cracking processes are very flexible and operating parameters can be adjusted to meet changing product demand. The three basic types of catalytic cracking processes are:

- Fluid catalytic cracking (FCC).

- Moving bed catalytic cracking.

- Thermofor catalytic cracking (TCC).

Fluid Catalytic Cracking

Fluid-bed catalytic crackers have a catalyst section (riser, reactor and regenerator) and a fractionating section, both operating together as an integrated processing unit. The FCC uses finely powdered catalyst, suspended in oil vapour or gas, which acts as a fluid. Cracking takes place in the feed pipe (riser) in which the mixture of catalyst and hydrocarbons flow through the reactor.

The FCC process mixes a preheated hydrocarbon charge with hot, regenerated catalyst as it enters the riser leading to the reactor. The charge combines with recycle oil within the riser, is vaporized and is raised to reactor temperature by the hot catalyst. As the mixture travels up the reactor, the charge is cracked at low pressure. This cracking continues until the oil vapours are separated from the catalyst in the reactor cyclones. The resultant product stream enters a column where it is separated into fractions, with some of the heavy oil directed back into the riser as recycle oil.

Spent catalyst is regenerated to remove coke which collects on the catalyst during the process. Spent catalyst flows through the catalyst stripper to the regenerator where it mixes with preheated air, burning off most of the coke deposits. Fresh catalyst is added and worn-out catalyst removed to optimize the cracking process.

Moving-bed Catalytic Cracking

Moving-bed catalytic cracking is similar to fluid catalytic cracking; however, the catalyst is in the form of pellets instead of fine powder. The pellets move continuously by conveyor or pneumatic lift tubes to a storage hopper at the top of the unit, and then flow downward by gravity through the reactor to a regenerator. The regenerator and hopper are isolated from the reactor by steam seals. The cracked product is separated into recycle gas, oil, clarified oil, distillate, naphtha and wet gas.

Thermofor Catalytic Cracking

In thermofor catalytic cracking, the preheated feedstock flows by gravity through the catalytic reactor bed. Vapours are separated from the catalyst and sent to a fractionating tower. The spent catalyst is regenerated, cooled and recycled, and the flue gas from regeneration is sent to a carbon monoxide boiler for heat recovery.

Health and Safety Considerations

Regular sampling and testing of feedstock, product and recycle streams should be performed to assure that the cracking process is working as intended and that no contaminants have entered the process stream. Corrosives or deposits in feedstock can foul gas compressors. When processing sour crude, corrosion may be expected where temperatures are below 482 °C. Corrosion takes place where both liquid and vapour phases exist and at areas subject to local cooling, such as nozzles and platform supports. When processing high-nitrogen feedstocks, exposure to ammonia and cyanide may subject carbon steel equipment in the FCC overhead system to corrosion, cracking or hydrogen blistering, which can be minimized by water wash or by corrosion inhibitors. Water wash may be used to protect overhead condensers in the main column subjected to fouling from ammonium hydrosulphide.

Critical equipment, including pumps, compressors, furnaces and heat exchangers should be inspected. Inspections should include checking for leaks due to erosion or other malfunctions such as catalyst buildup on the expanders, coking in the overhead feeder lines from feedstock residues, and other unusual operating conditions.

Liquid hydrocarbons in the catalyst or entering the heated combustion air stream can cause exothermic reactions. In some processes, caution must be taken to assure that explosive concentrations of catalyst dust are not present during recharge or disposal. When unloading coked catalyst, the possibility of iron sulphide fires exists. Iron sulphide will ignite spontaneously when exposed to air, and therefore needs to be wetted down with water to prevent it from becoming a source of ignition for vapours. Coked catalyst may either be cooled to below 49 °C before dumping from the reactor, or first dumped into containers purged with inert nitrogen and then cooled before further handling.

The possibility of exposure to extremely hot hydrocarbon liquids or vapours is present during process sampling or if a leak or release occurs. In addition, exposure to carcinogenic PAHs, aromatic naphtha containing benzene, sour gas (fuel gas from processes such as catalytic cracking and hydrotreating, which contains hydrogen sulphide and carbon dioxide), hydrogen sulphide and/or carbon monoxide gas may occur during a release of product or vapour. Inadvertent formation of highly toxic nickel carbonyl may occur in cracking processes that use nickel catalysts with resultant potential for hazardous exposures.

Catalyst regeneration involves steam stripping and decoking, which results in potential exposure to fluid waste streams which may contain varying amounts of sour water, hydrocarbon, phenol, ammonia, hydrogen sulphide, mercaptan and other materials, depending upon the feedstocks, crudes and processes. Safe work practices and the use of appropriate personal protective equipment (PPE) are needed when handling spent catalyst, recharging catalyst, or if leaks or releases occur.

Hydrocracking Process

Hydrocracking is a two-stage process combining catalytic cracking and hydrogenation, wherein distillate fractions are cracked in the presence of hydrogen and special catalysts to produce more desirable products. Hydrocracking has an advantage over catalytic cracking in that high-sulphur feedstocks can be processed without previous desulphurization. In the process, heavy aromatic feedstock is converted into lighter products under very high pressures and fairly high temperatures. When the feedstock has a high paraffinic content, the hydrogen prevents the formation of PAHs, reduces tar formation and prevents build-up of coke on the catalyst. Hydrocracking produces relatively large amounts of isobutane for alkylation feedstocks and also causes isomerization for pour point control and smoke point control, both of which are important in high-quality jet fuel.

In the first stage, feedstock is mixed with recycled hydrogen, heated and sent to the primary reactor, where a large amount of the feedstock is converted to middle distillates. Sulphur and nitrogen compounds are converted by a catalyst in the primary stage reactor to hydrogen sulphide and ammonia. The residual is heated and sent to a high-pressure separator, where hydrogen-rich gases are removed and recycled. The remaining hydrocarbons are stripped or purified to remove the hydrogen sulphide, ammonia and light gases, which are collected in an accumulator, where gasoline is separated from sour gas.

The stripped liquid hydrocarbons from the primary reactor are mixed with hydrogen and sent to the second-stage reactor, where they are cracked into high-quality gasoline, jet fuel and distillate blending stocks. These products go through a series of high- and low-pressure separators to remove gases, which are recycled. The liquid hydrocarbons are stabilized, split and stripped, with the light naphtha products from the hydrocracker used to blend gasoline while the heavier naphthas are recycled or sent to a catalytic reformer unit.

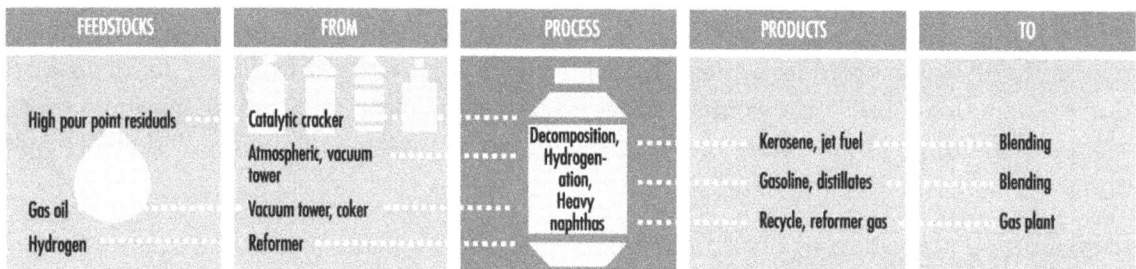

FEEDSTOCKS	FROM	PROCESS	PRODUCTS	TO
High pour point residuals	Catalytic cracker	Decomposition, Hydrogen-ation, Heavy naphthas	Kerosene, jet fuel	Blending
	Atmospheric, vacuum tower		Gasoline, distillates	Blending
Gas oil	Vacuum tower, coker		Recycle, reformer gas	Gas plant
Hydrogen	Reformer			

Hydrocracking process.

Health and Safety Considerations

Inspection and testing of safety relief devices are important due to the very high pressures in this process. Proper process control is needed to protect against plugging reactor beds. Because of the operating temperatures and presence of hydrogen, the hydrogen sulphide content of the feedstock must be strictly kept to a minimum in order to reduce the possibility of severe corrosion. Corrosion by wet carbon dioxide in areas of condensation must also be considered. When processing high-nitrogen feedstocks, the ammonia and hydrogen sulphide form ammonium hydrosulphide, which causes serious corrosion at temperatures below the water dew point. Ammonium hydrosulphide is also present in sour water stripping. Because the hydrocracker operates at very high pressures and temperatures, control of both hydrocarbon leaks and hydrogen releases is important to prevent fires.

Because this is a closed process, exposures are minimal under normal operating conditions. There is a potential for exposure to aliphatic naphtha containing benzene, carcinogenic PAHs, hydrocarbon gas and vapour emissions, hydrogen-rich gas and hydrogen sulphide gas as a result of high-pressure leaks. Large quantities of carbon monoxide may be released during catalyst regeneration and changeover. Catalyst steam stripping and regeneration creates waste streams containing sour water and ammonia. Safe work practices and appropriate personal protective equipment are needed when handling spent catalyst. In some processes, care is needed to assure that explosive concentrations of catalytic dust do not form during recharging. Unloading coked catalyst requires special precautions to prevent iron sulphideinduced fires. The coked catalyst should either be cooled to below 49 °C before dumping, or placed in nitrogen-inerted containers until cooled.

Combining Processes

Two combining processes, polymerization and alkylation, are used to join together small hydrogen-deficient molecules, called olefins, recovered from thermal and catalytic cracking, in order to create more desirable gasoline blending stocks.

Polymerization

Polymerization is the process of combining two or more unsaturated organic molecules (olefins) to form a single, heavier molecule with the same elements in the same proportion as the original molecule. It converts gaseous olefins, such as ethylene, propylene and butylene converted by thermal and fluid cracking units, into heavier, more complex, higher-octane molecules, including naphtha and petrochemical feedstocks. The olefin feedstock is pretreated to remove sulphur compounds and other undesirables, and then passed over a phosphorus catalyst, usually a solid catalyst or liquid phosphoric acid, where an exothermic polymeric reaction occurs. This requires the use of cooling water and the injection of cold feedstock into the reactor to control temperatures at various pressures. Acid in the liquids is removed by caustic wash, the liquids are fractionated, and the acid catalyst is recycled. The vapour is fractionated to remove butanes and neutralized to remove traces of acid.

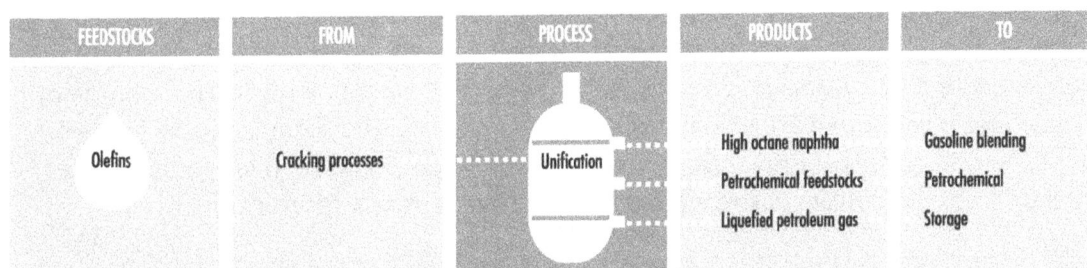

FEEDSTOCKS	FROM	PROCESS	PRODUCTS	TO
Olefins	Cracking processes	Unification	High octane naphtha / Petrochemical feedstocks / Liquefied petroleum gas	Gasoline blending / Petrochemical / Storage

Polymerization process.

Severe corrosion, leading to equipment failure, will occur should water contact the phosphoric acid, such as during water washing at shutdowns. Corrosion may also occur in piping manifolds, reboilers, exchangers and other locations where acid may settle out. There is a potential for exposure to caustic wash (sodium hydroxide), to phosphoric acid used in the process or washed out during turnarounds, and to catalyst dust. The potential for an uncontrolled exothermic reaction exists should loss of cooling water occur.

Alkylation

Alkylation combines the molecules of olefins produced from catalytic cracking with those of isoparaffins in order to increase the volume and octane of gasoline blends. Olefins will react with isoparaffins in the presence of a highly active catalyst, usually sulphuric acid or hydrofluoric acid (or aluminium chloride) to create a long-branched-chain paraffinic molecule, called *alkylate* (iso-octane), with exceptional anti-knock quality. The alkylate is then separated and fractionated. The relatively low reaction temperatures of 10 °C to 16 °C for sulphuric acid, 27 °C to 0 °C for hydrofluoric acid (HF) and 0°C for aluminium chloride, are controlled and maintained by refrigeration.

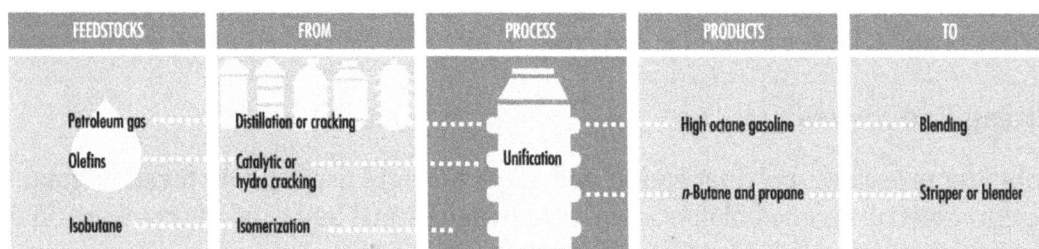

FEEDSTOCKS	FROM	PROCESS	PRODUCTS	TO
Petroleum gas	Distillation or cracking		High octane gasoline	Blending
Olefins	Catalytic or hydro cracking	Unification		
Isobutane	Isomerization		n-Butane and propane	Stripper or blender

Alkylation process.

Sulphuric acid alkylation: In cascade-type sulphuric acid alkylation units, feedstocks, including propylene, butylene, amylene and fresh isobutane, enter the reactor, where they contact the sulphuric acid catalyst. The reactor is divided into zones, with olefins fed through distributors to each zone, and the sulphuric acid and isobutanes flowing over baffles from zone to zone. Reaction heat is removed by evaporation of isobutane. The isobutane gas is removed from the top of the reactor, cooled and recycled, with a portion directed to the depropanizer tower. Residual from the reactor is settled, and the sulphuric acid is removed from the bottom of the vessel and recirculated. Caustic and/or water scrubbers are used to remove small amounts of acid from the process stream, which then goes to a de-isobutanizer tower. The debutanizer isobutane overhead is recycled, and the remaining hydrocarbons are separated in a rerun tower and/or sent to blending.

Hydrofluoric acid alkylation: There are two types of hydrofluoric acid alkylation processes: Phillips and UOP. In the Phillips process, olefin and isobutane feedstock is dried and fed to a combination reactor/settler unit. The hydrocarbon from the settling zone is charged to the main fractionator. The main fractionator overhead goes to a depropanizer. Propane, with trace amounts of hydrofluoric acid (HF), goes to an HF stripper, and is then catalytically defluorinated, treated and sent to storage. Isobutane is withdrawn from the main fractionator and recycled to the reactor/settler, and alkylate from the bottom of the main fractionator is sent to a splitter.

The UOP process uses two reactors with separate settlers. Half of the dried feedstock is charged to the first reactor, along with recycle and make-up isobutane, and then to its settler, where the acid is recycled and the hydrocarbon charged to the second reactor. The other half of the feedstock goes to the second reactor, with the settler acid being recycled and the hydrocarbons charged to the main fractionator. Subsequent processing is similar to Phillips in that the overhead from the main fractionator goes to a depropanizer, isobutane is recycled and alkylate is sent to a splitter.

Health and Safety Considerations

Sulphuric acid and hydrofluoric acid are dangerous chemicals, and care during delivery and unloading of acid is essential. There is a need to maintain sulphuric acid concentrations of 85 to 95% for good operation and to minimize corrosion. To prevent corrosion from hydrofluoric acid, acid concentrations inside the process unit must be maintained above 65% and moisture below 4%. Some corrosion and fouling in sulphuric acid units occurs from the breakdown of sulphuric acid esters, or where caustic is added for neutralization. These esters can be removed by fresh-acid treating and hot-water washing.

Upsets can be caused by loss of the coolant water needed to maintain process temperatures. Pressure on the cooling water and steam side of exchangers should be kept below the minimum pressure on the acid service side to prevent water contamination. Vents can be routed to soda ash scrubbers to neutralize hydrogen fluoride gas or hydrofluoric acid vapours before release. Curbs, drainage and isolation may be provided for process unit containment so that effluent can be neutralized before release to the sewer system.

Hydrofluoric acid units should be thoroughly drained and chemically cleaned prior to turnarounds and entry, to remove all traces of iron fluoride and hydrofluoric acid. Following shutdown, where water has been used, the unit should be thoroughly dried before hydrofluoric acid is introduced. Leaks, spills or releases involving hydrofluoric acid, or hydrocarbons containing hydrofluoric acid, are extremely hazardous. Precautions are necessary to assure that equipment and materials which have been in contact with acid are handled carefully and are thoroughly cleaned before they leave the process area or refinery. Immersion wash vats are often provided for neutralization of equipment which has come into contact with hydrofluoric acid.

There is a potential for serious hazardous and toxic exposures should leaks, spills or releases occur. Direct contact with sulphuric or hydrofluoric acid will cause severe skin and eye damage, and inhalation of acid mists or hydrocarbon vapours containing acid will cause severe irritation and damage to the respiratory system. Special precautionary emergency preparedness measures should be used, and protection should be provided that is appropriate to the potential hazard and areas possibly affected. Safe work practices and appropriate skin and respiratory personal protective equipment are needed where potential exposures to hydrofluoric and sulphuric acids during normal operations exist, such as reading gauges, inspecting and process sampling, as well as during emergency response, maintenance and turnaround activities. Procedures should be in place to assure that protective equipment and clothing worn in sulphuric or hydrofluoric acid activities, including chemical protective suits, head and shoe coverings, gloves, face and eye protection and respiratory protective equipment, are thoroughly cleaned and decontaminated before reissue.

Rearranging Processes

Catalytic reforming and isomerization are processes which rearrange hydrocarbon molecules to produce products with different characteristics. After cracking, some gasoline streams, although of the correct molecular size, require further processing to improve their performance, because they are deficient in some qualities, such as octane number or sulphur content. Hydrogen (steam) reforming produces additional hydrogen for use in hydrogenation processing.

Catalytic Reforming

Catalytic reforming processes convert low-octane heavy naphthas into aromatic hydrocarbons for petrochemical feedstocks and high-octane gasoline components, called reformates, by molecular rearrangement or dehydrogenation. Depending on the feedstock and catalysts, reformates can be produced with very high concentrations of toluene, benzene, xylene and other aromatics useful in gasoline blending and petrochemical processing. Hydrogen, a significant by-product, is separated from the reformate for recycling and use in other processes. The resultant product depends on reactor temperature and pressure, the catalyst used and the hydrogen recycle rate. Some catalytic reformers operate at low pressure and others at high pressure. Some catalytic reforming systems continuously regenerate the catalyst, some facilities regenerate all of the reactors during turn-arounds, and others take one reactor at a time off stream for catalyst regeneration.

In catalytic reforming, naphtha feedstock is pretreated with hydrogen to remove contaminants such as chlorine, sulphur and nitrogen compounds, which could poison the catalyst. The product is flashed and fractionated in towers where the remaining contaminants and gases are removed. The desulphurized naphtha feedstock is sent to the catalytic reformer, where it is heated to a vapour and passed through a reactor with a stationary bed of bi-metallic or metallic catalyst containing a small amount of platinum, molybdenum, rhenium or other noble metals. The two primary reactions which occur are production of high-octane aromatics by removing hydrogen from the feedstock molecules, and the conversion of normal paraffins to branched-chain or isoparaffins.

In platforming, another catalytic reforming process, feedstock which has not been hydrodesulphurized is combined with recycle gas and first passed over a less expensive catalyst. Any remaining impurities are converted to hydrogen sulphide and ammonia, and removed before the stream passes over the platinum catalyst. Hydrogen-rich vapour is recirculated to inhibit reactions which may poison the catalyst. The reactor output is separated into liquid reformate, which is sent to a stripping tower, and gas, which is compressed and recycled.

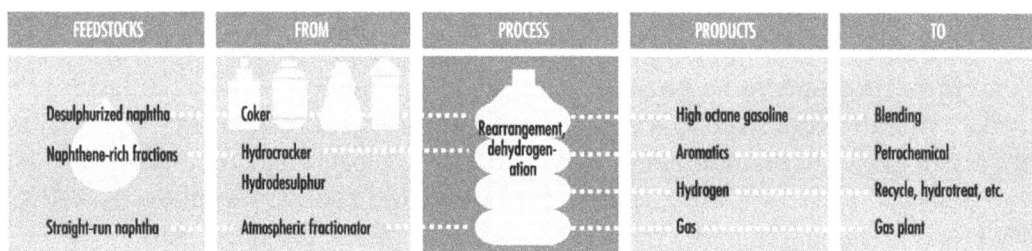

Catalytic reforming process.

Operating procedures are needed to control hot spots during start-up. Care must be taken not to break or crush the catalyst when loading the beds, as small fines will plug up the reformer screens. Precautions against dust when regenerating or replacing catalyst are needed. Small emissions of carbon monoxide and hydrogen sulphide may occur during regeneration of catalyst.

Water wash should be considered where stabilizer fouling has occurred in reformers due to the formation of ammonium chloride and iron salts. Ammonium chloride may form in pretreater exchangers and cause corrosion and fouling. Hydrogen chloride, from the hydrogenation of chlorine compounds, may form acids or ammonium chloride salt. The potential exists for exposure to aliphatic and aromatic naphthas, hydrogen-rich process gas, hydrogen sulphide and benzene should a leak or release occur.

Isomerization

Isomerization converts n-butane, n-pentane and n-hexane into their respective iso-paraffins. Some of the normal straight-chain paraffin components of light straight-run naphtha are low in octane. These can be converted to high-octane, branched-chain isomers by rearranging the bonds between atoms, without changing the number or kinds of atoms. Isomerization is similar to catalytic reforming in that the hydrocarbon molecules are rearranged, but unlike catalytic reforming, isomerization just converts normal paraffins to iso-paraffins. Isomerization uses a different catalyst than catalytic reforming.

The two distinct isomerization processes are butane (C_4) and pentane/hexane (C_5/C_6).

Butane (C_4) isomerization produces feedstock for alkylation. A lower-temperature process uses highly active aluminium chloride or hydrogen chloride catalyst without fired heaters, to isomerize *n*-butane. The treated and preheated feedstock is added to the recycle stream, mixed with HCl and passed through the reactor.

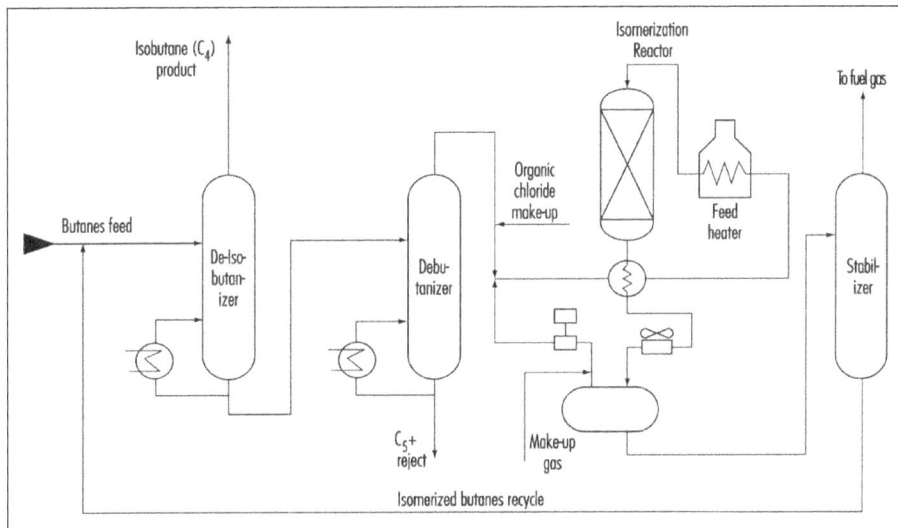

C4 isomerization.

Pentane/hexane isomerization is used to increase the octane number by converting *n*-pentane and *n*-hexane. In a typical pentane/hexane isomerization process, dried and desulphurized feedstock is mixed with a small amount of organic chloride and recycled hydrogen, and heated to reactor temperature. It is then passed over supported-metal catalyst in the first reactor, where benzene and olefins are hydrogenated. The feed next goes to the isomerization reactor, where the paraffins are catalytically isomerized to isoparaffins, cooled and passed to a separator. Separator gas and hydrogen, with make-up hydrogen, is recycled. The liquid is neutralized with alkaline materials and sent to a stripper column, where hydrogen chloride is recovered and recycled.

FEEDSTOCKS	FROM	PROCESS	PRODUCTS	TO
n-Butane			Isobutane	Alkylation
n-Pentane	Various processes	Rearrange-ment	Isopentane	Blending
n-Hexane			Isohexane	Blending
			Gas	Gas plant

Isomerization process.

If the feedstock is not completely dried and desulphurized, the potential exists for acid formation, leading to catalyst poisoning and metal corrosion. Water or steam must not be allowed to enter areas where hydrogen chloride is present. Precautions are needed to prevent HCl from entering sewers and drains. There is a potential for exposure to isopentane and aliphatic naphtha vapours and liquid, as well as to hydrogen-rich process gas, hydrochloric acid and hydrogen chloride, and to dust when solid catalyst is used.

Hydrogen Production (Steam Reforming)

High-purity hydrogen (95 to 99%) is needed for hydrodesulphurization, hydrogenation, hydroc-racking and petrochemical processes. If not enough hydrogen is produced as by-products of re-finery processes to meet the total refinery demand, the manufacture of additional hydrogen is required.

In hydrogen steam reforming, desulphurized gases are mixed with superheated steam and re-formed in tubes containing a nickel base catalyst. The reformed gas, which consists of steam, hy-drogen, carbon monoxide and carbon dioxide, is cooled and passed through converters where the carbon monoxide reacts with steam to form hydrogen and carbon dioxide. The carbon dioxide is scrubbed with amine solutions and vented to the atmosphere when the solutions are reactivated by heating. Any carbon monoxide remaining in the product stream is converted to methane.

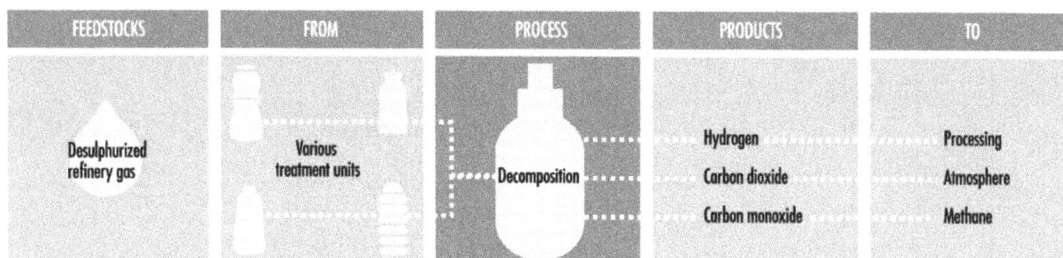

FEEDSTOCKS	FROM	PROCESS	PRODUCTS	TO
Desulphurized refinery gas	Various treatment units	Decomposition	Hydrogen / Carbon dioxide / Carbon monoxide	Processing / Atmosphere / Methane

Steam reforming process.

Inspections and testing must be conducted where the possibility exists for valve failure due to con-taminants in the hydrogen. Carryover from caustic scrubbers to prevent corrosion in preheaters must be controlled and chlorides from the feedstock or steam system prevented from entering reformer tubes and contaminating the catalyst. Exposures can result from contamination of con-densate by process materials such as caustics and amine compounds, and from excess hydrogen, carbon monoxide and carbon dioxide. The potential exists for burns from hot gases and superheat-ed steam should a release occur.

Miscellaneous Refinery Processes

Lubricant Base Stock and Wax Processes

Lubricating oils and waxes are refined from various fractions of atmospheric and vacuum distil-lation. With the invention of vacuum distillation, it was discovered that the waxy residuum made a better lubricant than any of the animal fats that were then in use, which was the beginning of modern hydrocarbon lubricant refining technology, whose primary objective is to remove unde-sirable products, such as asphalts, sulphonated aromatics and paraffinic and iso-paraffinic waxes from the residual fractions in order to produce high-quality lubricants. This is done by a series of processes including de-asphalting, solvent extraction and separation and treatment processes such as dewaxing and hydrofinishing.

In extraction processing, reduced crude from the vacuum unit is propane de-asphalted and com-bined with straight-run lubricating-oil feedstock, preheated and solvent extracted to produce a feedstock called raffinate. In a typical extraction process which uses phenol as the solvent, the feedstock is mixed with phenol in the treating section at temperatures below 204 °C. Phenol is then separated from the raffinate and recycled. The raffinate may then be subjected to another extraction process which uses furfural to separate aromatic compounds from non-aromatic hydrocarbons, producing a lighter-coloured raffinate with improved viscosity index and oxidation and thermal stability.

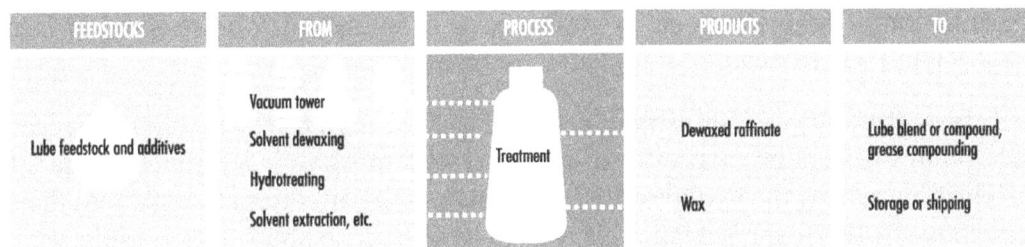

FEEDSTOCKS	FROM	PROCESS	PRODUCTS	TO
Lube feedstock and additives	Vacuum tower Solvent dewaxing Hydrotreating Solvent extraction, etc.	Treatment	Dewaxed raffinate Wax	Lube blend or compound, grease compounding Storage or shipping

Lubricating oil & wax manufacturing process.

Dewaxed raffinate may also be subject to further processing to improve the qualities of the base stock. Clay adsorbents are used to remove dark-coloured, unstable molecules from lubricating-oil base stocks. An alternate process, lube hydrofinishing, passes hot dewaxed raffinate and hydrogen through a catalyst that slightly changes the molecular structure, resulting in a lighter-coloured oil with improved characteristics. The treated lube oil base stocks are then mixed and/or compounded with additives to meet the required physical and chemical characteristics of motor oils, industrial lubricants and metal-working oils.

The two distinct types of wax derived from crude oil are paraffin wax, produced from distillate stocks, and microcrystalline wax, manufactured from residual stocks. Raffinate from the extraction unit contains a considerable amount of wax, which can be removed by solvent extraction and crystallization. The raffinate is mixed with a solvent, such as propane, methyl ethyl ketone (MEK) and toluene mixture or methyl isobutyl ketone (MIBK), and precooled in heat exchangers. The crystallization temperature is attained by the evaporation of the propane in the chiller and filter feed tanks. The wax is continuously removed by filters and cold solvent washed to recover retained oil. The solvent is recovered from the dewaxed raffinate by flashing and steam stripping, and recycled.

The wax is heated with hot solvent, chilled, filtered and given a final wash to remove all traces of oil. Before the wax is used, it may be hydro-finished to improve its odour and eliminate all traces of aromatics so the wax can be used in food processing. The dewaxed raffinate, which contains small amounts of paraffins, naphthenes and some aromatics, may be further processed for use as lubricating-oil base stocks.

Control of treater temperature is important to prevent corrosion from phenol. Wax can clog sewer or oil drainage systems and interfere with waste water treatment. The potential exists for exposure to process solvents such as phenol, propane, a methyl ethyl ketone and toluene mixture or methyl isobutyl ketone. Inhalation of hydrocarbon gases and vapours, aromatic naphtha containing benzene, hydrogen sulphide and hydrogen-rich process gas is a hazard.

Asphalt Processing

After primary distillation operations, asphalt is a portion of residual matter which requires further processing to impart characteristics required by its final use. Asphalt for roofing materials is produced by air blowing. Residual is heated in a pipe still almost up to its flashpoint and charged to a blowing tower where hot air is injected for a predetermined period of time. The dehydrogen ation of the asphalt forms hydrogen sulphide, and the oxidation creates sulphur dioxide. Steam is used to blanket the top of the tower to entrain the contaminants, and is passed through a scrubber to condense the hydrocarbons.

Vacuum distillation is generally used to produce road tar asphalt. The residual is heated and charged to a column where vacuum is applied to prevent cracking.

Condensed steam from the various asphalt processes will contain trace amounts of hydrocarbons. Any disruption of the vacuum can result in the entry of atmospheric air and subsequent fire. In asphalt production, raising the temperature of the vacuum tower bottom to improve efficiency can generate methane by thermal cracking. This creates vapours in asphalt storage tanks which are in the flammable range, but not detectable by flash testing. Air blowing can create some polynuclear aromatics (i.e., PAHs). Condensed steam from the air blowing asphalt process may also contain various contaminants.

Hydrocarbon Sweetening and Treating Processes

Many products, such as thermal naphthas from visbreaking, coking or thermal cracking, and high-sulphur naphthas and distillates from crude-oil distillation, require treating in order to be used in gasoline and fuel oil blends. Distillation products, including kerosene and other distillates, may contain trace amounts of aromatics, and naphthenes and lubricating-oil base stocks may contain wax. These undesirables are removed either at intermediate refining stages or just prior to sending products to blending and storage, by refining processes such as solvent extraction and solvent dewaxing. A variety of intermediate and finished products, including middle distillates, gasoline, kerosene, jet fuel and sour gases need to be dried and sweetened.

Treating is performed either at an intermediate stage in the refining process or just before sending finished products to blending and storage. Treating removes contaminants from oil, such as organic compounds containing sulphur, nitrogen and oxygen, dissolved metals, inorganic salts and soluble salts dissolved in emulsified water. Treating materials include acids, solvents, alkalis and oxidizing and adsorption agents. Acid treatments are used to improve the odour, colour and other properties of lube base stocks, to prevent corrosion and catalyst contamination, and to improve product stability. Hydrogen sulphide which is removed from "dry" sour gas by an absorbing agent (diethanolamine) is flared, used as a fuel or converted to sulphur. The type of treatment and agents depends on the crude feedstock, intermediate processes and end-product specifications.

Solvent Treatment Processes

Solvent extraction separates aromatics, naphthenes and impurities from product streams by dissolving or precipitation. Solvent extraction prevents corrosion, protects catalyst in subsequent processes and improves finished products by removing unsaturated, aromatic hydrocarbons from lubricant and grease base stocks.

The feedstock is dried and subjected to continuous countercurrent solvent treatment. In one process, feedstock is washed with a liquid in which the substances to be removed are more soluble than in the desired resultant product. In another process, selected solvents are added, causing impurities to precipitate out of the product. The solvent is separated from the product stream by heating, evaporation or fractionation, with residual trace amounts subsequently removed from the raffinate by steam stripping or vacuum flashing. Electric precipitation may be used for separation of inorganic compounds. The solvent is then regenerated to be used again in the process.

Typical chemicals used in the extraction process include a wide variety of acids, alkalis and solvents, including phenol and furfural, as well as oxidizing agents and adsorption agents. In the adsorption process, highly porous solid materials collect liquid molecules on their surfaces. The selection of specific processes and chemical agents depends on the nature of the feedstock being treated, the contaminants present and the finished product requirements.

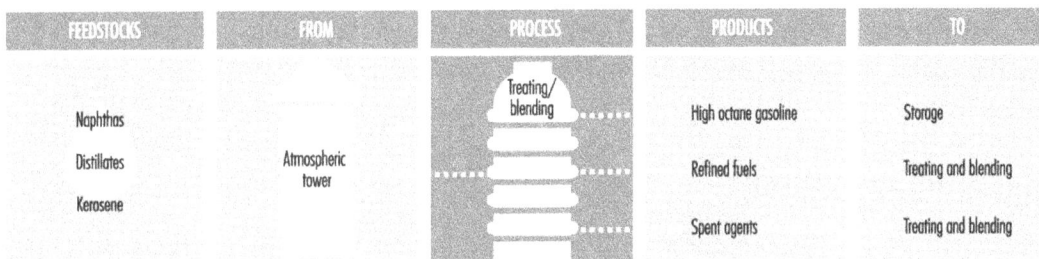

Solvent extraction process.

Solvent dewaxing removes wax from either distillate or residual base stocks, and may be applied at any stage in the refining process. In solvent dewaxing, waxy feedstocks are chilled by heat exchanger and refrigeration, and solvent is added to help develop crystals that are removed by vacuum filtration. The dewaxed oil and solvent are flashed and stripped, and the wax passes through a water settler, solvent fractionator and flash tower.

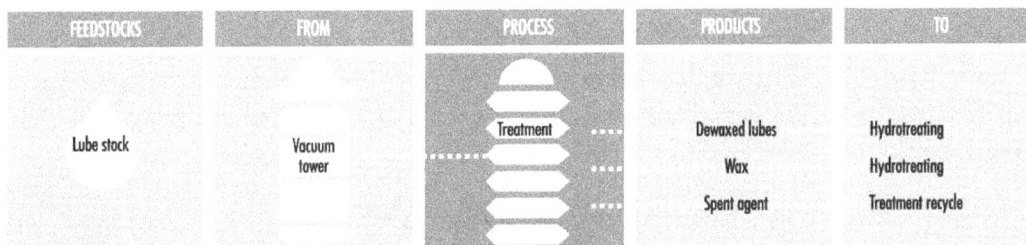

Solvent dewaxing process.

Solvent de-asphalting separates heavy oil fractions to produce heavy lubricating oil, catalytic cracking feedstock and asphalt. Feedstock and liquid propane (or hexane) are pumped to an extraction tower at precisely controlled mixtures, temperatures and pressures. Separation occurs in a rotating-disc contactor, based on differences in solubility. The products are then evaporated and steam stripped to recover propane for recycle. Solvent de-asphalting also removes sulphur and nitrogen compounds, metals, carbon residues and paraffins from feedstock.

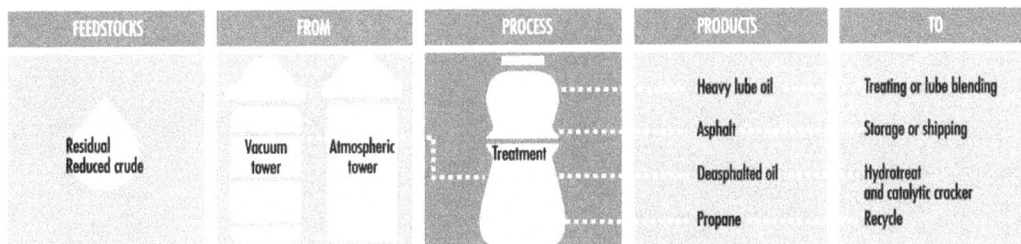

Solvent de-asphalting process.

Health and Safety Considerations

In solvent dewaxing, disruption of the vacuum will create a potential fire hazard by allowing air to enter the unit. The potential exists for exposure to dewaxing solvent vapours, a mixture of MEK and toluene. Although solvent extraction is a closed process, there is potential exposure to carcinogenic PAHs in the process oils and to extraction solvents such as phenol, furfural, glycol, MEK, amines and other process chemicals during handling and operations.

De-asphalting requires exact temperature and pressure control to avoid upset. In addition, moisture, excess solvent or a drop in operating temperature may cause foaming which affects the product temperature control and may create an upset. Contact with hot oil streams will cause skin burns. The potential exists for exposure to hot oil streams containing carcinogenic polycyclic aromatic compounds, liquefied propane and propane vapours, hydrogen sulphide and sulphur dioxide.

Hydrotreating Processes

Hydrotreating is used to remove about 90% of contaminants, including nitrogen, sulphur, metals and unsaturated hydrocarbons (olefins), from liquid petroleum fractions such as straight-run gasoline. Hydrotreating is similar to hydrocracking in that both the hydrogen and the catalyst are used to enrich the hydrogen content of the olefin feedstock. However, the degree of saturation is not as great as that achieved in hydrocracking. Typically, hydrotreating is done prior to processes such as catalytic reforming, so that the catalyst is not contaminated by untreated feedstock. Hydrotreating is also used before catalytic cracking to reduce sulphur and improve product yields, and to upgrade middle distillate petroleum fractions into finished kerosene, diesel fuel and heating fuel oils.

Hydrotreating processes differ depending upon the feedstocks and catalysts. Hydrodesulphurization removes sulphur from kerosene, reduces aromatics and gum-forming characteristics, and saturates any olefins. Hydroforming is a dehydrogenation process used to recover excess hydrogen and produce high-octane gasoline. Hydrotreated products are blended or used as catalytic reforming feedstock.

In catalytic hydrodesulphurization, the feedstock is de-aerated, mixed with hydrogen, preheated and charged under high pressure through a fixed-bed catalytic reactor. The hydrogen is separated and recycled and the product stabilized in a stripper column where the light ends are removed.

During this process, sulphur and nitrogen compounds present in the feedstock are converted to hydrogen sulphide (H_2S) and ammonia (NH_3). Residual hydrogen sulphide and ammonia are removed either by steam stripping, by a combination high- and low-pressure separator or by amine wash which recovers hydrogen sulphide in a highly concentrated stream suitable for conversion into elemental sulphur.

FEEDSTOCKS	FROM	PROCESS	PRODUCTS	TO
Naphthas, distillates sour gas oil, residuals	Atmospheric and vacuum tower	Treating, hydrogen-ation	Naphtha	Catalytic reformer
			Hydrogen	Recycle
	Catalytic and thermal cracker		Distillates	Blending
			H_2S, ammonia	Sulphur plant, treater
			Gas	Gas plant

Hydrodesulphurization process.

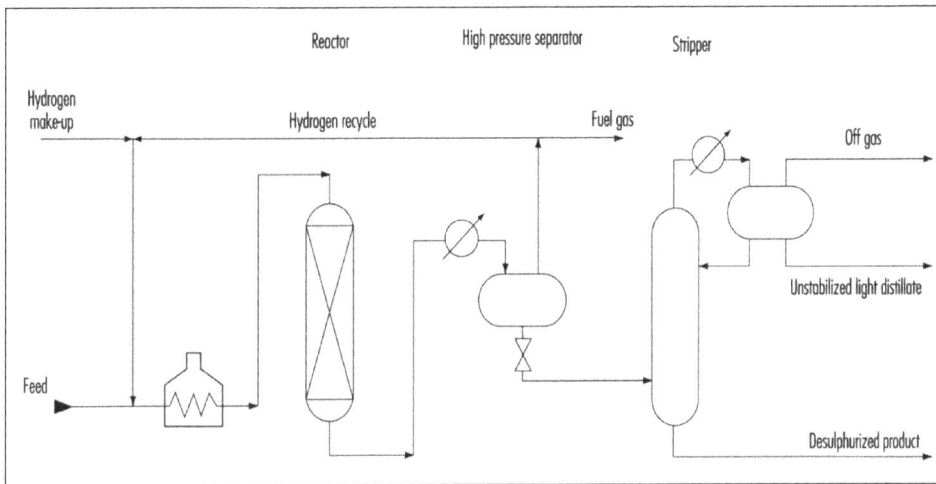

Schematic of hydrodesulphurization process.

In hydrotreating, the hydrogen sulphide content of the feedstock must be strictly controlled to a minimum to reduce corrosion. Hydrogen chloride may form and condense as hydrochloric acid in the lower-temperature portions of the unit. Ammonium hydrosulphide may form in high-temperature, high-pressure units. In the event of a release, there is a potential for exposure to aromatic naphtha vapours which contain benzene, hydrogen sulphide or hydrogen gas, or to ammonia should a sour water leak or spill occur. Phenol may also be present if high-boiling-point feedstocks are processed.

Excessive contact time and/or temperature will create coking in the unit. Precautions need to be taken when unloading coked catalyst from the unit to prevent iron sulphide fires. The coked catalyst should be cooled to below 49 °C before removal, or dumped into nitrogen-inerted bins where it can be cooled before further handling. Special anti-foam additives may be used to prevent catalyst poisoning from silicone carryover in coker feedstock.

Other Sweetening and Treating Processes

Treatment, drying and sweetening processes are used to remove impurities from blending stocks.

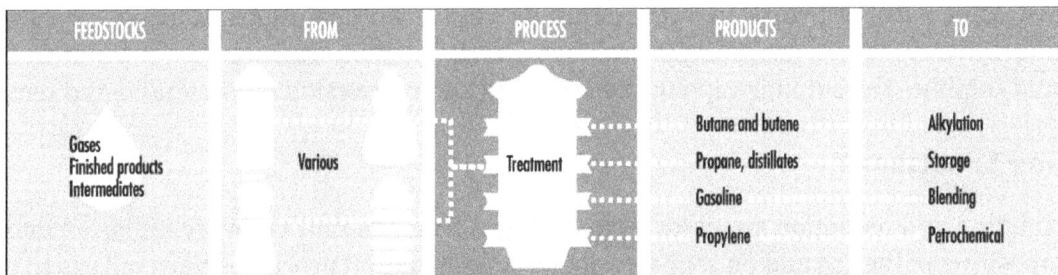

FEEDSTOCKS	FROM	PROCESS	PRODUCTS	TO
Gases Finished products Intermediates	Various	Treatment	Butane and butene Propane, distillates Gasoline Propylene	Alkylation Storage Blending Petrochemical

Sweetening and treating processes.

Sweetening processes use air or oxygen. If excess oxygen enters these processes, it is possible for a fire to occur in the settler due to the generation of static electricity. There is a potential for exposure to hydrogen sulphide, sulphur dioxide, caustic (sodium hydroxide), spent caustic, spent catalyst (Merox), catalyst dust and sweetening agents (sodium carbonate and sodium bicarbonate).

Amine Acid Gas Treatment Plants

Sour gas (fuel gas from processes such as catalytic cracking and hydrotreating, which contains hydrogen sulphide and carbon dioxide) must be treated before it can be used as refinery fuel. Amine plants remove acid contaminants from sour gas and hydrocarbon streams. In amine plants, gas and liquid hydrocarbon streams containing carbon dioxide and/or hydrogen sulphide are charged to a gas absorption tower or liquid contactor, where the acid contaminants are absorbed by counterflowing amine solutions—monoethanolamine (MEA), diethanolamine (DEA) or methyldiethanolamine (MDEA). The stripped gas or liquid is removed overhead, and the amine is sent to a regenerator. In the regenerator, the acidic components are stripped by heat and reboiling action, and disposed of, while the amine is recycled.

In order to minimize corrosion, proper operating practices should be established, and regenerator bottom and reboiler temperatures need to be controlled. Oxygen should be kept out of the system to prevent amine oxidation. There is potential for exposure to amine compounds (i.e., MEA, DEA, MDEA), hydrogen sulphide and carbon dioxide.

Sweetening and Drying

Sweetening (mercaptan removal) treats sulphur compounds (hydrogen sulphide, thiophene and mercaptan) to improve colour, odour and oxidation stability, and reduces concentrations of carbon dioxide in gasoline. Some mercaptans are removed by having the product make contact with water-soluble chemicals (e.g., sulphuric acid) that react with the mercaptans. Caustic liquid (sodium hydroxide), amine compounds (diethanolamine) or fixed-bed catalyst sweetening may be used to convert mercaptans to less objectionable disulphides.

Product drying (water removal) is accomplished by water absorption, with or without adsorption agents. Some processes simultaneously dry and sweeten by adsorption on molecular sieves.

Sulphur Recovery

Sulphur recovery removes hydrogen sulphide from sour gases and hydrocarbon streams. The Clause process converts hydrogen sulphide to elemental sulphur through the use of thermal and catalytic reactions. After burning hydrogen sulphide under controlled conditions, knockout pots remove water and hydrocarbons from feed-gas streams, which are then exposed to a catalyst to recover additional sulphur. The sulphur vapour from burning and conversion is condensed and recovered.

Tail Gas Treatment

Both oxidation and reduction are used to treat tail gas from sulphur recovery units, depending on the composition of the gas and on refinery economics. Oxidation processes burn tail gas to convert all sulphur compounds to sulphur dioxide, and reduction processes convert sulphur compounds to hydrogen sulphide.

Hydrogen Sulphide Scrubbing

Hydrogen sulphide scrubbing is a primary hydrocarbon feedstock treating process used to prevent catalyst poisoning. Depending on the feedstock and the nature of the contaminants, desulphurization methods will vary from ambient-temperature-activated charcoal absorption to high-temperature catalytic hydrogenation followed by zinc oxide treating.

Sat and Unsat Gas Plants

Feedstocks from various refinery units are sent to gas treating plants, where butanes and butenes are removed for use as alkylation feedstock, heavier components are sent to gasoline blending, propane is recovered for LPG and propylene is removed for use in petrochemicals.

Sat gas plants separate components from refinery gases, including butanes for alkylation, pentanes for gasoline blending, LPGs for fuel and ethane for petrochemicals. There are two different sat gas processes: absorption-fractionation or straight fractionation. In absorption-fractionation, gases and liquids from various units are fed to an absorber/de-ethanizer where C2 and lighter fractions are separated by lean-oil absorption and removed for use as fuel gas or petrochemical feed. The remaining heavier fractions are stripped and sent to a debutanizer, and the lean oil is recycled back to the absorber/de-ethanizer. C3/C4 is separated from pentanes in the debutanizer, scrubbed to remove hydrogen sulphide, and fed to a splitter to separate propane and butane. The absorption stage is eliminated in fractionation plants. Sat gas processes depend on feedstock and product demand.

Corrosion occurs from the presence of hydrogen sulphide, carbon dioxide and other compounds as a result of prior treating. Streams containing ammonia should be dried before processing. Anti-fouling additives are used in absorption oil to protect heat exchangers. Corrosion inhibitors are used to control corrosion in overhead systems. The potential exists for exposure to hydrogen sulphide, carbon dioxide, sodium hydroxide, MEA, DEA and MDEA to be carried over from prior treating.

Unsat gas plants recover light hydrocarbons from wet gas streams from catalytic crackers and delayed coker overhead accumulators or fractionation receivers. In a typical process, wet gases are compressed and treated with amine to remove hydrogen sulphide either before or after entering a fractionating absorber, where they mix into a concurrent flow of debutanized gasoline. The light fractions are separated by heat in a reboiler, with the offgas sent to a sponge absorber and the bottoms sent to a debutanizer. A portion of the debutanized hydrocarbon is recycled, and the balance goes to a splitter for separation. Overhead gases go to a depropanizer for use as alkylation unit feedstock.

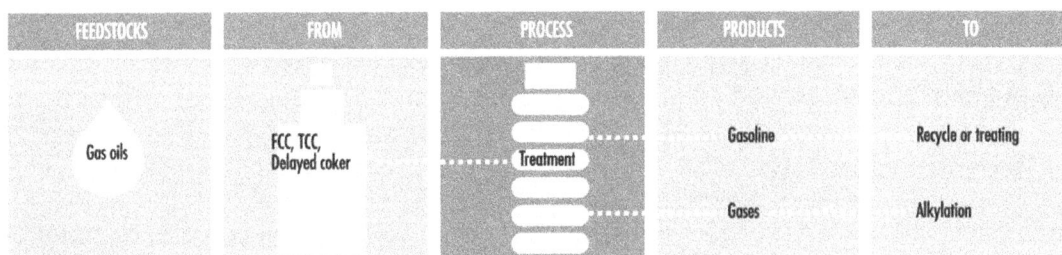

Unsat gas plant process.

Corrosion can occur from moist hydrogen sulphide and cyanides in unsat gas plants which handle FCC feedstocks. Corrosion from hydrogen sulphide and deposits in the high-pressure sections of gas compressors from ammonium compounds is possible when feedstocks are from the delayed coker or the TCC. The potential exists for exposure to hydrogen sulphide and to amine compounds such as MEA, DEA and MDEA.

Gasoline, Distillate Fuel and Lubricant Base Stock Blending Processes

Blending is the physical mixture of a number of different liquid hydrocarbon fractions to produce finished products with specific desired characteristics. Products can be blended in-line through a manifold system or batch blended in tanks and vessels. In-line blending of gasoline, distillates, jet fuel and lubricant base stocks is accomplished by injecting proportionate amounts of each component into the main stream where turbulence promotes thorough mixing:

- Gasolines are blends of reformates, alkylates, straight-run gasoline, thermal and catalytically cracked gasolines, coker gasoline, butane and appropriate additives.

- Fuel oil and diesel fuel are blends of distillates and cycle oils, and jet fuel may be straight-run distillate or blended with naphtha.

- Lubricating oils are blends of refined base stocks.

- Asphalt is blended from various residual stocks depending on its intended use.

Additives are often mixed into gasoline and motor fuels during or after blending to provide specific properties not inherent in petroleum hydrocarbons. These additives include octane enhancers, anti-knock agents, anti-oxidants, gum inhibitors, foam inhibitors, rust inhibitors, carburettor (carbon) cleaners, detergents for injector cleaning, diesel odourizers, colour dyes, distillate anti-static, gasoline oxidizers such as methanol, ethanol and methyl tertiary butyl ether, metal deactivators and others.

Batch and in-line blending operations require strict controls to maintain desired product quality. Spills should be cleaned and leaks repaired to avoid slips and falls. Additives in drums and bags need to be handled properly to avoid strain and exposure. The potential for contacting hazardous additives, chemicals, benzene and other materials exists during blending, and appropriate engineering controls, personal protective equipment and proper hygiene are needed to minimize exposures.

Auxiliary Refinery Operations

Auxiliary operations supporting refinery processes include those which provide process heat and cooling; provide pressure relief; control air emissions; collect and treat waste water; provide utilities such as power, steam, air and plant gases; and pump, store, treat and cool process water.

Waste Water Treatment

Refinery waste water includes condensed steam, stripping water, spent caustic solutions, cooling tower and boiler blowdown, wash water, alkaline and acid waste neutralization water and other process-associated water. Waste water typically contains hydrocarbons, dissolved materials, suspended solids, phenols, ammonia, sulphides and other compounds. Waste water treatment is used for process water, runoff water and sewerage water prior to their discharge. These treatments may require permits, or there must be recycling.

The potential exists for fire should vapours from waste water containing hydrocarbons reach a source of ignition during the treatment process. The potential exists for exposure to the various chemicals and waste products during process sampling, inspection, maintenance and turnarounds.

Pretreatment

Pretreatment is the initial separation of hydrocarbons and solids from waste water. API separators, interceptor plates and settling ponds are used to remove suspended hydrocarbons, oily sludge and solids by gravity separation, skimming and filtration. Acidic waste water is neutralized with ammonia, lime or soda ash. Alkaline waste water is treated with sulphuric acid, hydrochloric acid, carbon dioxide-rich flue gas or sulphur. Some oil-in-water emulsions are first heated to help separate the oil and the water. Gravity separation depends on the different specific gravities of water and immiscible oil globules, which allows free oil to be skimmed off the surface of the waste water.

Sour Water Stripping

Water containing sulphides, called sour water, is produced in catalytic cracking and hydro-treating processes, and whenever steam is condensed in the presence of gases containing hydrogen sulphide.

Stripping is used on waste water containing sulphides and/or ammonia, and solvent extraction is used to remove phenols from waste water. Waste water which is to be recycled may require cooling to remove heat and/or oxidation by spraying or air stripping to remove any remaining phenols, nitrates and ammonia.

Secondary Treatment

Following pretreatment, suspended solids are removed by sedimentation or air flotation. Waste water with low levels of solids is screened or filtered, and flocculation agents may be added to help separation. Materials with high adsorption characteristics are used in fixed-bed filters or added to the waste water to form a slurry which is removed by sedimentation or filtration. Secondary treatment processes biologically degrade and oxidize soluble organic matter by the use of activated sludge, unaerated or aerated lagoons, trickling filter methods or anaerobic treatments. Additional treatment methods are used to remove oils and chemicals from waste water.

Tertiary Treatment

Tertiary treatments remove specific pollutants in order to meet regulatory discharge requirements. These treatments include chlorination, ozonation, ion exchange, reverse osmosis, activated carbon adsorption, and others. Compressed oxygen may be diffused into waste water streams to oxidize certain chemicals or to satisfy regulatory oxygen content requirements.

Cooling Towers

Cooling towers remove heat from process water by evaporation and latent heat transfer between hot water and air. The two types of towers are counterflow and crossflow:

- In counterflow cooling, hot process water is pumped to the uppermost plenum and allowed to fall through the tower. Numerous slats, or spray nozzles, are located throughout the length of the tower to disperse the water flow and help in cooling. Simultaneously, air enters at the tower bottom, creating a concurrent flow of air against the water. Induced draft towers have the fans at the air outlet. Forced draft towers have the fans or blowers at the air inlet.

- Crossflow towers introduce airflow at right angles to the water flow throughout the structure.

Recirculated cooling water must be treated to remove impurities and any dissolved hydrocarbons. Impurities in cooling water can corrode and foul piping and heat exchangers, scale from dissolved salts can deposit on pipes, and wooden cooling towers can be damaged by micro-organisms.

Cooling tower water can be contaminated by process materials and by-products, including sulphur dioxide, hydrogen sulphide and carbon dioxide, with resultant exposures. There is potential for exposure to water treatment chemicals or to hydrogen sulphide when waste water is treated in conjunction with cooling towers. Because the water is saturated with oxygen from being cooled with air, the chances for corrosion are intensified. One means of corrosion prevention is the addition of a material to the cooling water which forms a protective film on pipes and other metal surfaces.

When cooling water is contaminated by hydrocarbons, flammable vapours can evaporate into the discharge air. If a source of ignition or lightning is present, fires may start. Fire hazards exist when there are relatively dry areas in induced-draft cooling towers of combustible construction. Loss of power to cooling tower fans or water pumps can create serious consequences in process operations.

Steam Generation

Steam is produced through heater and boiler operations in central steam generation plants and at various process units, using heat from flue gas or other sources. Steam generation systems include:

- Heaters (furnaces), with their burners and a combustion air system.

- Draft or pressure systems to remove flue gas from the furnace, soot blowers, and compressed air systems which seal openings to prevent flue gas from escaping.

- Boilers, consisting of a number of tubes which carry the water/steam mixture through the furnace providing for maximum heat transfer (these tubes run between steam distribution drums at the top of the boiler, and water collecting drums at the bottom of the boiler).

- Steam drums to collect steam and direct it to the superheater before it enters the steam distribution system.

The most potentially hazardous operation in steam generation is heater start-up. A flammable mixture of gas and air can build up as a result of loss of flame at one or more burners during light-off. Specific start-up procedures are required for each different type of unit, including purging before light-off and emergency procedures in the event of misfire or loss of burner flame. If feedwater runs low and boilers are dry, the tubes will overheat and fail. Excess water will be carried over into the steam distribution system, causing damage to the turbines. Boilers should have continuous or intermittent blowdown systems to remove water from steam drums and to limit build-up of scale on turbine blades and superheater tubes. Care must be taken not to overheat the superheater during start-up and shut down. Alternate fuel sources should be provided in event of loss of fuel gas due to refinery unit shutdown or emergency.

Heater Fuel

Any one or any combination of fuels, including refinery gas, natural gas, fuel oil and powdered coal may be used in heaters. Refinery off-gas is collected from process units and combined with natural gas and LPG in a fuel gas balance drum. The balance drum provides constant system pressure, fairly stable BTU (energy) content fuel and automatic separation of suspended liquids in gas vapours, and prevents carryover of large slugs of condensate into the distribution system.

Fuel oil is typically a mix of refinery crude oil and straight-run and cracked residues, blended with other products. The fuel oil system delivers fuel to process unit heaters and steam generators at required temperatures and pressures. The fuel oil is heated to pumping temperature, sucked through a coarse suction strainer, pumped to a temperature-control heater and then through a fine mesh strainer before being burned. Knockout pots, provided at process units, are used to remove liquids from fuel gas before burning.

In one example of process unit heat generation, carbon monoxide (CO) boilers recover heat in catalytic cracking units as carbon monoxide in flue gas is burned to complete combustion. In other processes, waste heat recovery units use heat from the flue gas to make steam.

Steam Distribution

Steam typically is generated by heaters and boilers combined into one unit. Steam leaves the boilers at the highest pressure required by the process units or the electrical generator. The steam pressure is then reduced in turbines which drive process pumps and compressors. When refinery steam is also used to drive steam turbine generators to produce electricity, the steam must be produced at much higher pressure than required for process steam. The steam distribution system consists of valves, fittings, piping and connections which are suitable for the pressure of the steam transported. Most steam used in the refinery is condensed to water in heat exchangers and reused as boiler feedwater, or discharged to waste water treatment.

Steam Feedwater

Feedwater supply is an important part of steam generation. There must always be as many pounds of water entering the steam generation system as there are pounds of steam leaving it. Water used in steam generation must be free of contaminants, including minerals and dissolved impurities, which can damage the system or affect the operation. Suspended materials such as silt, sewage and oil, which form scale and sludge, are coagulated or filtered out of the water. Dissolved gases, particularly carbon dioxide and oxygen which cause boiler corrosion, are removed by de-aeration and treatment. Dissolved minerals such as metallic salts, calcium and carbonates, which cause scale, corrosion and turbine blade deposits, are treated with lime or soda ash to precipitate them out of the water. Depending on its characteristics, raw boiler feedwater may be treated by clarification, sedimentation, filtration, ion exchange, de-aeration and internal treatment. Recirculated cooling water must also be treated to remove hydrocarbons and other contaminants.

Process Heaters, Heat Exchangers and Coolers

Process heaters and heat exchangers preheat feedstocks in distillation towers and in refinery processes to reaction temperatures. The major portion of heat provided to process units comes from fired heaters found on crude and reformer preheater units, coker heaters and large-column reboilers, which are fueled by refinery or natural gas, distillate and residual oils. Heaters are usually designed for specific process operations, and most are either cylindrical vertical or box-type designs. Heat exchangers use either steam or hot hydrocarbon, transferred from some other section of the process, for heat input.

Heat is also removed from some processes by air and water exchangers, fin fans, gas and liquid coolers and overhead condensers, or by transferring the heat to other systems. The basic mechanical vapour compression refrigeration system is designed to serve one or more process units, and includes an evaporator, compressor, condenser, controls and piping. Common coolants are water, alcohol/water mixture or various glycol solutions.

A means of providing adequate draft or steam purging is required to reduce the chance of explosions when lighting fires in heater furnaces. Specific start-up and emergency procedures are required for each type of unit. If fire impinges on fin fans, failure could occur due to overheating. If flammable product escapes from a heat exchanger or cooler due to a leak, a fire could occur.

Care must be taken to assure that all pressure is removed from heater tubes before removing any header or fitting plugs. Consideration should be given to providing for pressure relief in heat exchanger piping systems in the event they are blocked off while full of liquid. If controls fail, variations of temperature and pressure could occur on either side of the heat exchanger. If heat exchanger tubes fail and process pressure is greater than heater pressure, product could enter the heater with downstream consequences. If the pressure is less, the heater stream could enter into the process fluid stream. If loss of circulation occurs in liquid or gas coolers, increased product temperature could affect downstream operations, requiring pressure relief.

Depending on the fuel, process operation and unit design, there is a potential for exposure to hydrogen sulphide, carbon monoxide, hydrocarbons, steam boiler feedwater sludge and water treatment chemicals. Skin contact with boiler blowdown which may contain phenolic compounds should be avoided. Exposure to radiant heat, superheated steam and hot hydrocarbons is possible.

Pressure Relief and Flare Systems

Engineering controls which are incorporated into processes include reducing flammable vapour concentrations by ventilation, dilution and inerting. Pressurization is used to maintain control rooms above atmospheric pressure in order to reduce the possibility of vapours entering. Pressure relief systems are provided to control vapours and liquids which are released by pressure-relieving devices and blowdowns. Pressure relief is an automatic, planned release when operating pressure reaches a predetermined level. Blowdown usually refers to the intentional release of material, such as blowdowns from process unit start-ups, furnace blowdowns, shutdowns and emergencies. Vapour depressuring is the rapid removal of vapours from pressure vessels in case of emergency. This may be accomplished by the use of a rupture disc, usually set at a higher pressure than the relief valve.

Safety Relief Valves

Safety relief valves, used to control air, steam, gas and hydrocarbon vapour and liquid pressures, open in proportion to the increase in pressure over the normal operating pressure. Safety valves, designed primarily to release high volumes of steam, usually pop open to full capacity. The over-pressure needed to open liquid relief valves, where large-volume discharge is not required, increases as the valve lifts due to increased spring resistance. Pilot-operated safety release valves, with up to six times the capacity of normal relief valves, are used where tighter sealing and larger-volume discharges are required. Non-volatile liquids are usually pumped to oil/water separation and recovery systems, and volatile liquids are sent to units operating at a lower pressure.

Flares

A typical closed pressure-release and flare system includes relief valves and lines from process units for collection of discharges, knockout drums to separate vapours and liquids, seals and/or purge gas for flashback protection and a flare and igniter system, which combusts vapours if discharge direct to the atmosphere is not permitted. Steam may be injected into the flare tip to reduce visible smoke.

Liquids should not be allowed to discharge to a vapour disposal system. Flare knockout drums and flares need to be large enough to handle emergency blowdowns, and drums require relief in event of overpressure. Provide pressure relief valves where the potential exists for overpressure in refinery processes, such as due to the following causes:

- Loss of cooling water, possibly resulting in a greatly increased pressure drop in condensers, in turn increasing the pressure in the process unit.

- Rapid vaporization and pressure increase from injection of a lower-boiling-point liquid, including water, into a process vessel operating at higher temperatures.

- Expansion of vapour and resultant overpressure due to overheated process steam, malfunctioning heaters or fire.

- Failure of automatic controls, closed outlets, heat exchanger failure, etc.

- Internal explosion, chemical reaction, thermal expansion, accumulated gases, etc.

- Loss of reflux, causing a pressure rise in distillation towers.

Because the quantity of reflux affects the volume of vapours leaving the distillation tower, loss of volume causes a pressure drop in condensers and a pressure rise in distillation towers.

Maintenance is important because valves are required to function properly. Common valve operating problems include:

- Failure to open at set pressure due to plugging of the valve inlet or outlet or by corrosion, preventing proper operation of the disc holder and guides.

- Failure to reseat after popping open due to fouling, corrosion or deposits on the seat or moving parts, or by solids in the gas stream cutting the valve disc.

- Chattering and premature opening, due to operating pressure being too close to the valve set point.

Utilities

Water: Depending on location and community resources, refineries may draw upon public water supplies for drinking and process water or may have to pump and treat their own potable water. Treatment may include a wide range of requirements, from desalting to filtration, chlorination and testing.

Sewage: Also, depending on availability of community or private offsite treatment plants, refineries may have to provide for the permitting, collection, treatment and discharge of their sanitary waste.

Electric power: Refineries either receive electricity from outside sources or produce their own, using electric generators driven by steam turbines or gas engines. Areas are classified with regard to the type of electrical protection required to prevent a spark from igniting vapours or contain an explosion within electrical equipment. Electrical substations, which are normally located in non-classified areas, away from sources of flammable hydrocarbon vapour or cooling tower water spray, contain transformers, circuit breakers and feed circuit switches. Substations feed power to distribution stations within the process unit areas. Distribution stations can be located in classified areas, provided that electrical classification requirements are met. Distribution stations typically use a liquid-filled transformer provided with an oil-filled or air-break disconnect device.

Normal electrical safety precautions, including dry footing, "high voltage" warning signs and guarding should be implemented to protect against electrocution. Employees should be familiar with refinery electrical safe work procedures. Lockout/tagout and other appropriate safe work practices should be implemented to prevent energizing while work is being performed on high-voltage electrical equipment. Hazardous exposures may occur when working around transformers and switches which contain a dielectric fluid requiring special handling precautions.

Turbine, Gas and Air Compressor Operations

Air and Gas Compressors

Refinery exhaust ventilation and air supply systems are designed to capture or dilute gases, fumes, dusts and vapours which may contaminate working spaces or the outside atmosphere. Captured contaminants are reclaimed if feasible, or directed to disposal systems after being cleaned or burned. Air supply systems include compressors, coolers, air receivers, air dryers, controls and distribution piping. Blowers are also used to provide air to certain processes. Plant air is provided for the operation of air-powered tools, catalyst regeneration, process heaters, steam-air decoking, sour water oxidation, gasoline sweetening, asphalt blowing and other uses. Instrument air is provided for use in pneumatic instruments and controls, air motors and purge connections. Plant gas, such as nitrogen, is provided for inerting vessels and other uses. Both reciprocating and centrifugal compressors are used for gas and compressed air.

Air compressors should be located so that the suction does not take in flammable vapours or corrosive gases. There is a potential for fire should a leak occur in gas compressors. Knockout drums are needed to prevent liquid surges from entering gas compressors. If gases are contaminated with solid materials, strainers are needed. Failure of automatic compressor controls will affect processes. If maximum pressure could potentially be greater than compressor or process equipment design pressure, pressure relief should be provided. Guarding is needed for exposed moving parts on compressors. Compressor buildings should be properly electrically classified, and provisions made for proper ventilation.

Where plant air is used as back-up to instrument air, interconnections must be upstream of the instrument air drying system to prevent contamination of instruments with moisture. Alternate sources of instrument air supply, such as use of nitrogen, may be needed in the event of power outages or compressor failure. Apply appropriate safeguards so that gas, plant air and instrument air are not used as the source for breathing or for pressuring potable water systems.

Turbines

Turbines are usually gas or steam powered and are used to drive pumps, compressors, blowers and other refinery process equipment. Steam enters turbines at high temperatures and pressures, expanding across and driving rotating blades while directed by fixed blades.

Steam turbines used for exhaust operating under vacuum need a safety relief valve on the discharge side for protection and to maintain steam in event of vacuum failure. Where maximum operating pressure could be greater than design pressure, steam turbines need relief devices. Consideration should be given to providing governors and overspeed-control devices on turbines.

Pumps, Piping and Valves

Centrifugal and positive displacement (reciprocating) pumps are used to move hydrocarbons, process water, fire water and waste water throughout the refinery. Pumps are driven by electric motors, steam turbines or internal combustion engines.

Process and utility piping systems distribute hydrocarbons, steam, water and other products throughout the facility. They are sized and constructed of materials dependent on the type of service, pressure, temperature and nature of the products. There are vent, drain and sample connections on piping, as well as provisions for blanking. Different types of valves, including gate valves, bypass valves, globe and ball valves, plug valves, block and bleed valves and check valves are used, depending on their operating purpose. These valves can be operated manually or automatically.

Valves and instrumentation which require servicing or other work should be accessible at grade level or from an operating platform. Remote-controlled valves, fire valves and isolation valves may be used to limit the loss of product at pump suction lines in the event of leakage or fire. Operating vent and drain connections may be provided with double block valves, or a block valve and plug or blind flange for protection against releases. Depending on the product and service, backflow prevention from the discharge line may be needed. Provisions may be made for pipeline expansion, movement and temperature changes to avoid rupture. Pumps operated with reduced or no flow can overheat and rupture. The failure of automatic pump controls could cause a deviation in process pressure and temperature. Pressure relief in the discharge piping should be provided where pumps can be overpressured.

Tank Storage

Atmospheric storage tanks and pressure storage tanks are used throughout the refinery for storage of crudes, intermediate hydrocarbons (those used for processing) and finished products, both liquids and gases. Tanks are also provided for fire water, process and treatment water, acids, air and hydrogen, additives and other chemicals. The type, construction, capacity and location of tanks depends on their use and the nature, vapour pressure, flashpoints and pour points of the materials stored. Many types of tanks are used in refineries, the simplest being above-ground, cone-roof tanks for storage of combustible (non-volatile) liquids such as diesel fuels, fuel oils and lubricating oils. Open-top and covered (internal) floating-roof tanks, which store flammable (volatile) liquids such as gasoline and crude oil, restrict the amount of space between the top of the product and the tank roof in order to maintain a vapour-rich atmosphere to preclude ignition.

The potential for fire exists if hydrocarbon storage tanks are overfilled or develop leaks which allow liquid and vapours to escape and reach sources of ignition. Refineries should establish manual gauging and product receipt procedures to control overfills or provide automatic overflow control and signaling systems on tanks. Tanks may be equipped with fixed or semi-fixed foam-water fire protection systems. Remote-controlled valves, isolation valves and fire valves may be provided at tanks for pump-out or closure in the event of a fire inside the tank or in the tank dike or storage area. Tank venting, cleaning and confined-space entry programmes are used to control work inside tanks, and hot work permit systems are used to control sources of ignition in and around storage tanks.

Handling, Shipping and Transportation

Loading gases and liquid hydrocarbons into pipelines, tank cars, tank trucks and marine vessels and barges for transport to terminals and consumers is the final refinery operation. Product characteristics, distribution needs, shipping requirements, fire prevention, and environmental protection and operating criteria are important when designing marine docks, loading racks and pipeline manifolds. Operating procedures need to be established and agreed to by the shipper and receiver, and communications maintained during product transfer. Tank trucks and rail tank cars may be either top or bottom loaded. Loading and unloading liquefied petroleum gas (LPG) requires special considerations over and above those for liquid hydrocarbons. Where required, vapour recovery systems should be provided at loading racks and marine docks.

Safe work practices and appropriate personal protective equipment may be needed when loading or unloading, cleaning up spills or leaks, or when gauging, inspecting, sampling or performing maintenance activities on loading facilities or vapour recovery systems. Delivery should be stopped or diverted in the event of an emergency such as a tank truck or tank car compartment overfill.

A number of different hazardous and toxic chemicals are used in refineries, varying from small amounts of test reagents used in laboratories to large quantities of sulphuric acid and hydrofluoric acids used in alkaline processing. These chemicals need to be received, stored and handled properly. Chemical manufacturers provide material safety information which can be used by refineries to develop safety procedures, engineering controls, personal protection requirements and emergency response procedures for handling chemicals.

The nature of the hazard at loading and unloading facilities depends upon the products being loaded and the products previously transported in the tank car, tank truck or marine vessel. Bonding equalizes the electrical charge between the loading rack and the tank truck or tank car. Grounding prevents the flow of stray currents at truck and rail loading facilities. Insulating flanges are used on marine dock piping connections to prevent static electricity build-up and discharges. Flame arrestors are installed in loading rack and marine vapour recovery lines to prevent flashback. Where switch loading is permitted, safe procedures should be established and followed.

Automatic or manual shutoff systems at supply headers should be provided at top- and bottom-loading racks and marine docks in the event of leaks or overfills. Anti-fall protection, such as hand rails, may be needed for docks and top-loading racks. Drainage and recovery systems may be provided at loading racks for storm drainage, at docks and to handle spills and leaks. Precautions are needed at LPG-loading facilities so as not to overload or overpressurize tank cars and trucks.

Refinery Support Activities and Facilities

A number of different facilities, activities and programmes, each of which has its own specific safety and health requirements, are needed to support refinery processes depending on the refinery's location and available resources.

Administrative Activities

A wide variety of administrative support activities, depending on the refining company's philosophy and the availability of community services, are required to assure continued operation of a refinery. The function which controls oil movements into, within and out from the refinery is unique to refineries. The administrative functions can be broken down as follows. The day-to-day operation of the process units is the operations function. Another function is responsible for assuring that arrangements have been made for a continuous supply of crude oil. Other functional activities include medical services (both emergency and continuing health care), food service, engineering services, janitorial services and routine administrative and management functions common to most industries, such as accounting, purchasing, human relations and so on. The refinery training function is responsible for supervisor and employee skills and crafts training including initial, refresher and remedial training, and for employee and contractor orientation and training in emergency response and safe work practices and procedures.

Construction and Maintenance

The continued safe operation of refineries depends upon the establishment and implementation of programmes and procedures for regular maintenance and preventive maintenance, and assuring replacement when necessary. Turnarounds, wherein the entire refinery or entire process units will be shut down for total equipment overall and replacement at one time, is a type of preventive maintenance programme unique to the process industry. Mechanical integrity activities, such as inspection, repair, testing and certification of valves and relief devices, which are part of the process safety management programme, are important to the continued safe operation of a refinery, as are maintenance work orders for the continued effectiveness of the refinery "management of change" programme. Work permit programmes control hot work and safe work, such as isolation and lockout, and entry into confined spaces. Maintenance and instrumentation shops have purposes which include:

- Delicate and precise work to test, maintain and calibrate refinery process controls, instruments and computers.

- Welding.

- Equipment repair and overhaul.

- Vehicle maintenance.

- Carpentry and so on.

Construction and maintenance safety and health relies on some of the following programmes.

Isolation

The safe maintenance, repair and replacement of equipment within process units often requires the isolation of tanks, vessels and lines in order to preclude the possibility of flammable liquids or vapours entering an area where hot work is being performed. Isolation is normally attained by disconnecting and closing off all of the piping leading to or from a vessel; blinding or blanking the pipe at a connection near the tank or vessel; or closing a double set of block valves on the piping, if provided, and opening a bleeder valve between the two closed valves.

Lockout/Tagout

Lockout and tagout programmes prevent the inadvertent activation of electrical, mechanical, hydraulic or pneumatically energized equipment during repair or maintenance. All electrically powered equipment should have its circuit breaker or main switch locked or tagged out and tested to assure non-operability, prior to starting work. Mechanical hydraulic and pneumatic equipment should be de-energized and have its power source locked or tagged out prior to starting work. Valve closing lines which are being worked on, or which are isolated, should also be locked out or tagged to prevent unauthorized opening.

Metallurgy

Metallurgy is used to assure the continued strength and integrity of lines, vessels, tanks and reactors which are subject to corrosion from the acids, corrosives, sour water, and gases and other chemicals created by and used in processing crude oil. Non-destructive testing methods are employed throughout the refinery to detect excessive corrosion and wear before failure occurs. Proper safety precautions are required to prevent excessive exposures to workers who are handling or are exposed to radioactive testing equipment, dyes and chemicals.

Warehouses

Warehouses store not only the parts, materials and equipment needed for continued refinery operations, but also store packaged chemicals and additives that are used in maintenance, processing and blending. Warehouses may also maintain supplies of required personal protective clothing and equipment including hard hats, gloves, aprons, eye and face protection, respiratory protection, safety and impervious footwear, flame-resistant clothing and acid-protective clothing. Proper storage and separation of flammable and combustible liquids and hazardous chemicals is needed to prevent spills, fires and mixing of incompatible products.

Safety and Environmental and Occupational Hygiene

Other important refinery support activities are safety, fire prevention and protection, environmental protection and industrial hygiene. These may be provided as separate functions or integrated into the refinery operations. Safety, emergency preparedness and response, and fire prevention and protection activities are often the responsibility of the same function within a refinery.

The safety function participates in process safety management programmes as part of the design review, pre-construction and construction review and pre-start-up review teams. Safety often assists in the contractor qualification process, reviews contractor activities and investigates incidents involving employees and contractors. Safety personnel may be responsible for overseeing permit-required activities such as confined space entry and hot work, and for checking the availability and readiness of portable fire extinguishers, decontamination facilities, safety showers, eye wash stations, fixed detection devices and alarms, and emergency self-contained breathing apparatus placed at strategic locations in event of a toxic gas release.

Safety programmes: The refinery safety function usually has responsibility for the development and administration of various safety and incident prevention programmes, including, but not limited to, the following:

- Design construction and pre-start-up safety reviews.
- Accident, incident and near miss investigation and reporting.
- Emergency preparedness plans and response programmes.
- Contractor safety programme.
- Safe work practices and procedures.
- Lockout/tagout.
- Confined and inert space entry.
- Scaffolding.
- Electrical safety, equipment grounding and fault protection programme.
- Machine guarding.
- Safety signs and notices.
- Hot work, safe work and entry permit systems.

Fire brigades: Refinery fire brigades and emergency responders may be full-time brigade members; designated refinery employees, such as operators and maintenance personnel who are trained and assigned to respond in addition to their regular duties; or a combination of both. Besides fires, brigades traditionally respond to other refinery incidents such as acid or gas releases, rescue from vessels or tanks, spills and so on. The fire protection function may be responsible for the inspection and testing of fire detectors and signals, and fixed and portable fire protection systems and equipment, including fire trucks, fire pumps, fire water lines, hydrants, hoses and nozzles.

Refinery firefighting differs from normal firefighting because rather than extinguishment, it is often preferable to allow certain fires to continue to burn. In addition, each type of hydrocarbon liquid, gas and vapour has unique fire chemistry characteristics which must be thoroughly understood in order to best control their fires. For example, extinguishment of a hydrocarbon vapour fire without first stopping the vapour release, would only create a continued vapour gas cloud with the probability of re-ignition and explosion. Fires in tanks containing crude oil and heavy residuals need to be handled with specific firefighting techniques to avoid the possibility of an explosion or tank boil-over.

Hydrocarbon fires are often extinguished by stopping the flow of product and allowing the fire to burn out while applying cooling water to protect adjacent equipment, tanks and vessels from heat exposures. Many fixed fire protection systems are designed with this specific purpose. Fighting fires in process units under pressure requires special consideration and training, particularly when catalysts such as hydrofluoric acid are involved. Special firefighting chemicals, such as dry powder and foam-water solutions, may be used to extinguish hydrocarbon fires and control vapour emissions.

Emergency preparedness: Refineries need to develop and implement emergency response plans for a number of different potential situations, including explosions, fires, releases and rescues. The emergency plans should include the use of outside assistance, including contractors, governmental and mutual aid as well as availability of special supplies and equipment, such as firefighting foam and spill containment and adsorption materials.

Gas and Vapour Testing

Gas, particulate and vapour monitoring, sampling and testing in refineries is conducted to assure that work can be performed safely and processes can be operated without toxic or hazardous exposures, explosions or fires. Atmospheric testing is conducted using a variety of instruments and techniques to measure oxygen content, hydrocarbon vapours and gases, and to determine hazardous and toxic exposure levels. Instruments must be properly calibrated and adjusted prior to use, by qualified persons, to assure dependable and accurate measurements. Depending on the work location, potential hazards and type of work being performed, testing, sampling and monitoring may be conducted prior to the start of work, or at specified intervals during work, or continuously throughout the course of work.

When establishing refinery procedures for sampling and testing flammable, inert and toxic atmospheres, the use of personal protective equipment, including appropriate respiratory protection, should be considered. It should be noted that canister-type respirators are unsuitable for oxygen-deficient atmospheres. Testing requirements should depend upon the degree of hazard which would be present in the event of instrument failure.

Testing of the following substances may be performed using portable equipment or fixed instrumentation:

Oxygen: Combustible gas meters work by burning a minute sample of the atmosphere being tested. In order to obtain an accurate combustible gas reading, a minimum of 10% and a maximum of 25% oxygen must be present in the atmosphere. The amount of oxygen present in the atmosphere is determined by using an oxygen meter prior to, or simultaneously with, using the combustible gas meter. Testing for oxygen is essential when working in confined or enclosed spaces, as entry without respiratory protection (provided that there are no toxic exposures) requires normal breathing-air oxygen concentrations of approximately 21%. Oxygen meters are also used to measure the amount of oxygen present in inerted spaces, to assure that there is not enough present to support combustion during hot work or other operations.

Hydrocarbon vapours and gases: "Hot work" is work which creates a source of ignition, such as welding, cutting, grinding, blast cleaning, operating an internal combustion engine and so on, in an area where the potential for exposure to flammable vapours and gases exists. In order to conduct hot work safely, instruments known as combustible gas meters are used to test the atmosphere for hydrocarbon vapours. Hydrocarbon vapours or gases will burn only when mixed with air (oxygen) in certain proportions and ignited. If there is not enough vapour in the air, the mixture is said to be "too lean to burn", and if there is too much vapour (too little oxygen), the mixture is "too rich to burn". The limiting proportions are called the "upper and lower flammable limits" and are expressed as a percentage of volume of vapour in air. Each hydrocarbon molecule or mixture has different flammability limits, typically ranging from about 1 to 10% vapour in air. Gasoline vapour, for example, has a lower flammable limit of 1.4% and an upper flammable limit of 7.6 per cent.

Toxic atmospheres: Special instruments are used to measure the levels of toxic and hazardous gases, vapours and particulates which may be present in the atmosphere where people are working. These measurements are used to determine the level and type of protection needed, which may vary from complete ventilation and replacement of the atmosphere to the use of respiratory and personal protective equipment by people working in the area. Examples of hazardous and toxic exposures which may be found in refineries include asbestos, benzene, hydrogen sulphide, chlorine, carbon dioxide, sulphuric and hydrofluoric acids, amines, phenol and others.

Health and Safety Programmes

The basis for refinery industrial hygiene is an administrative and engineering controls programme covering facility exposures to toxic and hazardous chemicals, laboratory safety and hygiene, ergonomics and medical surveillance.

Regulatory agencies and companies establish exposure limitations for various toxic and hazardous chemicals. The occupational hygiene function conducts monitoring and sampling to measure employee exposure to hazardous and toxic chemicals and substances. Industrial hygienists may develop or recommend engineering controls, preventive work practices, product substitution, personal protective clothing and equipment or alternate measures of protection or reducing exposure.

Medical programmes: Refineries typically require preplacement and periodic medical examinations to determine the employee's ability to initially and subsequently perform the work, and assure that the continued work requirements and exposures will not endanger the employee's health or safety.

Personal protection: Personal protection programmes should cover typical refinery exposures, such as noise, asbestos, insulation, hazardous waste, hydrogen sulphide, benzene and process chemicals including caustics, hydrogen fluoride, sulphuric acid and so on. Industrial hygiene may designate the appropriate personal protective equipment to be used for various exposures, including negative pressure and air-supplied respirators and hearing, eye and skin protection.

Product safety: Product safety awareness covers knowing about the hazards of chemicals and materials to which the potential for exposure exists in the workplace, and what actions to take in the event exposure by ingestion, inhalation or skin contact occurs. Toxicological studies of crude oil, refinery streams, process chemicals, finished products and proposed new products are conducted to determine the potential effects of exposure on both employees and consumers. The data are used to develop health information concerning permissible limits of exposure or acceptable amounts of hazardous materials in products. This information is typically distributed by material safety data sheets (MSDSs) or similar documents, and employees are trained or educated in the hazards of the materials in the workplace.

Environmental Protection

Environmental protection is an important consideration in refinery operations because of regulatory compliance requirements and a need for conservation as oil prices and costs escalate. Oil refineries produce a wide range of air and water emissions that can be hazardous to the environment. Some of these are contaminants in the original crude oil, while others are a result of refinery processes and operations. Air emissions include hydrogen sulphide, sulphur dioxide, nitrogen oxides and carbon monoxide. Waste water typically contains hydrocarbons, dissolved materials, suspended solids, phenols, ammonia, sulphides, acids, alkalis and other contaminants. There is also the risk of accidental spills and leaks of a wide range of flammable and/or toxic chemicals.

Controls established to contain liquid and vapour releases and reduce operating costs include the following:

- Energy conservation: Controls include steam leak control and condensate recovery programmes to conserve energy and increase efficiency.

- Water pollution: Controls include waste water treatment in API separators and subsequent treatment facilities, storm water collection, retainment and treatment and spill prevention containment and control programmes.

- Air pollution: Since refineries operate continuously, leak detection, particularly at valves and pipe connections, is important. Controls include reducing hydrocarbon vapour emissions and releases to the atmosphere, refinery valve and fitting tightness programmes, floating roof tank seals and vapour containment programmes, and vapour recovery for loading and unloading facilities and for venting tanks and vessels.

- Ground pollution: Preventing oil spillage from polluting soil and contaminating ground water is accomplished by the use of dikes and the providing of drainage to specified, protected containment areas. Contamination from spillage inside dike areas may be prevented by the use of secondary containment measures, such as impervious plastic or clay dike liners.

- Spill response: Refineries should develop and implement programmes to respond to spills of crude oil, chemicals and finished products, on both land and water. These programmes may rely on trained employees or outside agencies and contractors to respond to the emergency. The type, amount needed and availability of spill clean-up and restoration supplies and equipment, either on site or on call, should be included in the preparedness plan.

References

- Extraction, oil, world-industries: economywatch.com, Retrieved 24 March 2019

- Drilling: studentenergy.org, Retrieved 15 May 2019

- Field-development-phase, upstream: oil-gasportal.com, Retrieved 14 April 2019

- Oil-and-gas-development, seaoilmine, products: toyoeng.com, Retrieved 11 January 2019

- Petroleum-production-phase, upstream: oil-gasportal.com, Retrieved 09 June 2019

Chapter 6

Petroleum Products

There are numerous products which are extracted from petroleum, such as diesel fuel, gasoline/ petrol, paraffin wax, kerosene, petroleum coke, lubricating oil, liquefied petroleum gas, petroleum jelly, naphtha, etc. The topics elaborated in this chapter will help in gaining a better perspective about these petroleum products.

Petroleum products are materials derived from crude oil (petroleum) as it is processed in oil refineries. Unlike petrochemicals, which are a collection of well-defined usually pure chemical compounds, petroleum products are complex mixtures. The majority of petroleum is converted to petroleum products, which includes several classes of fuels.

Diesel Fuel

Diesel fuel, also called diesel oil, combustible liquid used as fuel for diesel engines, ordinarily obtained from fractions of crude oil that are less volatile than the fractions used in gasoline. In diesel engines the fuel is ignited not by a spark, as in gasoline engines, but by the heat of air compressed in the cylinder, with the fuel injected in a spray into the hot compressed air. Diesel fuel releases more energy on combustion than equal volumes of gasoline, so diesel engines generally produce better fuel economy than gasoline engines. In addition, the production of diesel fuel requires fewer refining steps than gasoline, so retail prices of diesel fuel traditionally have been lower than those of gasoline (depending on the location, season, and taxes and regulations). On the other hand, diesel fuel, at least as traditionally formulated, produces greater quantities of certain air pollutants such as sulfur and solid carbon particulates, and the extra refining steps and emission-control mechanisms put into place to reduce those emissions can act to reduce the price advantages of diesel over gasoline. In addition, diesel fuel emits more carbon dioxide per unit than gasoline, offsetting some of its efficiency benefits with its greenhouse gas emissions.

Several grades of diesel fuel are manufactured—for example, "light-middle" and "middle" distillates for high-speed engines with frequent and wide variations in load and speed (such as trucks and automobiles) and "heavy" distillates for low- and medium-speed engines with sustained loads and speeds (such as trains, ships, and stationary engines). Performance criteria are cetane number (a measure of ease of ignition), ease of volatilization, and sulfur content. The highest grades, for automobile and truck engines, are the most volatile, and the lowest grades, for low-speed engines, are the least volatile, leave the most carbon residue, and commonly have the highest sulfur content.

Sulfur is a critical polluting component of diesel and has been the object of much regulation. Traditional "regular" grades of diesel fuel contained as much as 5,000 parts per million (ppm) by weight sulfur. In the 1990s "low sulfur" grades containing no more than 500 ppm sulfur were introduced, and in the following years even lower levels of sulfur were required. Regulations in the United States required that by 2010 diesel fuels sold for highway vehicles be "ultra-low sulfur" (ULSD) grades, containing a maximum of 15 ppm. In the European Union, regulations required that from 2009 diesel fuel sold for road vehicles be only so-called "zero-sulfur" or "sulfur-free" diesels, containing no more than 10 ppm. Lower sulfur content reduces emissions of sulfur compounds implicated in acid rain and allows diesel vehicles to be equipped with highly effective emission-control systems that would otherwise be damaged by higher concentrations of sulfur. Heavier grades of diesel fuel, made for use by off-road vehicles, ships and boats, and stationary engines, are generally allowed higher sulfur content, though the trend has been to reduce limits in those grades as well.

In addition to traditional diesel fuel refined from petroleum, it is possible to produce so-called synthetic diesel, or Fischer-Tropsch diesel, from natural gas, from synthesis gas derived from coal, or from biogas obtained from biomass. Also, biodiesel, a biofuel, can be made primarily from oily plants such as the soybean or oil palm. These alternative diesel fuels can be blended with traditional diesel fuel or used alone in diesel engines without modification, and they have very low sulfur content. Alternative diesel fuels are often proposed as means to reduce dependence on petroleum and to reduce overall emissions, though only biodiesel can provide a life cycle carbon dioxide benefit.

Refining

Petroleum diesel, also called petrodiesel, or fossil diesel is produced from petroleum and is a hydrocarbon mixture, obtained in the fractional distillation of crude oil between 200 °C and 350 °C at atmospheric pressure.

A modern diesel pump.

Fuel Value and Price

The density of petroleum diesel is about 0.85 kg/l (7.09 lbs/gallon(us)), about 18 percent more than petrol (gasoline), which has a density of about 0.72 kg/l (6.01 lbs/gallon(us)). When burnt, diesel typically releases about 38.6 MJ/l (138,700 BTU per US gallon), whereas gasoline releases 34.9 MJ/l (125,000 BTU per US gallon), 10 percent less by energy density, but 45.41 MJ/kg and 48.47 MJ/kg, 6.7 percent more by specific energy. Diesel is generally simpler to refine from petroleum than gasoline. The price of diesel traditionally rises during colder months as demand for heating oil rises, which is refined in much the same way. Due to its higher level of pollutants, diesel must undergo additional filtration which contributes to a sometimes higher cost. In many parts of the United States and throughout the UK and Australia diesel may be higher priced than petrol. Reasons for higher priced diesel include the shutdown of some refineries in the Gulf of Mexico, diversion of mass refining capacity to gasoline production, and a recent transfer to ultra-low sulfur diesel (ULSD), which causes infrastructural complications.

Use as Vehicle Fuel

Unlike Petroleum ether and Liquefied petroleum gas engines, diesel engines do not use high voltage spark ignition (spark plugs). An engine running on diesel compresses the air inside the cylinder to high pressures and temperatures (compression ratios from 15:1 to 21:1 are common); the diesel is generally injected directly into the cylinder near the end of the compression stroke. The high temperatures inside the cylinder causes the diesel fuel to react with the oxygen in the mix (burn or oxidize), heating and expanding the burning mixture in order to convert the thermal/pressure difference into mechanical work; i.e., to move the piston. (Glow plugs are used to assist starting the engine to preheat cylinders to reach a minimum operating temperature.) High compression ratios and throttleless operation generally result in diesel engines being more efficient than many spark-ignited engines.

This and being less flammable and explosive than gasoline are the main reasons for military use of diesel in armored fighting vehicles like tanks and trucks. Engines running on diesel also provide more torque and are less likely to stall as they are controlled by a mechanical or electronic governor. A disadvantage of diesel as a vehicle fuel in some climates, compared to gasoline or other petroleum derived fuels, is that its viscosity increases quickly as the fuel's temperature decreases, turning into a non-flowing gel at temperatures as high as -19 °C (-2.2 °F) or -15 °C (+5 °F), which can't be pumped by regular fuel pumps. Special low temperature diesel contains additives that keep it in a more liquid state at lower temperatures, yet starting a diesel engine in very cold weather still poses considerable difficulties.

Another rare disadvantage of diesel engines compared to petrol/gasoline engines is the possibility of runaway failure. Since diesel engines do not require spark ignition, they can sustain operation as long as diesel fuel is supplied. Fuel is typically supplied via a fuel pump. If the pump breaks down in an "open" position, the supply of fuel will be unrestricted and the engine will runaway and risk terminal failure.

Use as Car Fuel

Diesel-powered cars generally have a better fuel economy than equivalent gasoline engines and produce less greenhouse gas emission. Their greater economy is due to the higher energy per-liter content of diesel fuel and the intrinsic efficiency of the diesel engine. While petrodiesel's higher density results in higher greenhouse gas emissions per liter compared to gasoline, the 20–40 percent better fuel economy achieved by modern diesel-engined automobiles offsets the higher-per-liter emissions of greenhouse gases, and produces 10-20 percent less greenhouse gas emissions than comparable gasoline vehicles. Biodiesel-powered diesel engines offer substantially improved emission reductions compared to petro-diesel or gasoline-powered engines, while retaining most of the fuel economy advantages over conventional gasoline-powered automobiles.

Reduction of Sulfur Emissions

In the past, diesel fuel contained higher quantities of sulfur. European emission standards and preferential taxation have forced oil refineries to dramatically reduce the level of sulfur in diesel fuels. In the United States, more stringent emission standards have been adopted with the transition to ULSD starting in 2006 and becoming mandatory on June 1, 2010. U.S. diesel fuel typically also has a lower cetane number (a measure of ignition quality) than European diesel, resulting in worse cold weather performance and some increase in emissions. This is one reason why U.S. drivers of large trucks have increasingly turned to biodiesel fuels with their generally higher cetane ratings.

Environment Hazards of Sulfur

High levels of sulfur in diesel are harmful for the environment because they prevent the use of catalytic diesel particulate filters to control diesel particulate emissions, as well as more advanced technologies, such as nitrogen oxide (NOx) adsorbers (still under development), to reduce emissions. However, the process for lowering sulfur also reduces the lubricity of the fuel, meaning that additives must be put into the fuel to help lubricate engines. Biodiesel and biodiesel/petrodiesel blends, with their higher lubricity levels, are increasingly being utilized as an alternative. The U.S. annual consumption of diesel fuel in 2006 was about 190 billion liters (42 billion imperial gallons or 50 billion US gallons).

Chemical Composition

Diesel is immiscible with water.

Petroleum-derived diesel is composed of about 75 percent saturated hydrocarbons (primarily paraffins including n, iso, and cycloparaffins), and 25 percent aromatic hydrocarbons (including naphthalenes and alkylbenzenes). The average chemical formula for common diesel fuel is $C_{12}H_{23}$, ranging from approx. $C_{10}H_{20}$ to $C_{15}H_{28}$.

Algae, Microbes and Water

Algae require sunlight to live and grow. As there is no sunlight in a closed fuel tank, no algae can survive there. However, some microbes can survive there, and can feed on the diesel fuel.

These microbes form a colony that lives at the fuel/water interface. They grow quite rapidly in warmer temperatures. They can even grow in cold weather when fuel tank heaters are installed. Parts of the colony can break off and clog the fuel lines and fuel filters.

It is possible to either kill this growth with a biocide treatment, or eliminate the water, a necessary component of microbial life. There are a number of biocides on the market, which must be handled very carefully. If a biocide is used, it must be added every time a tank is refilled until the problem is fully resolved.

Biocides attack the cell wall of microbes resulting in lysis, the death of a cell by bursting. The dead cells then gather on the bottom of the fuel tanks and form a sludge, filter clogging will continue after biocide treatment until the sludge has abated.

Given the right conditions microbes will repopulate the tanks and re-treatment with biocides will then be necessary. With repetitive biocide treatments microbes can then form resistance to a particular brand. Trying another brand may resolve this.

Road Hazard

Petrodiesel spilled on a road will stay there until washed away by sufficiently heavy rain, whereas gasoline will quickly evaporate. After the light fractions have evaporated, a greasy slick is left on the road which can destabilize moving vehicles. Diesel spills severely reduce tire grip and traction, and have been implicated in many accidents. The loss of traction is similar to that encountered on black ice. Diesel slicks are especially dangerous for two-wheeled vehicles such as motorbikes.

Synthetic Diesel

Wood, hemp, straw, corn, garbage, food scraps, and sewage-sludge may be dried and gasified to synthesis gas. After purification the Fischer-Tropsch process is used to produce synthetic diesel. This means that synthetic diesel oil may be one route to biomass based diesel oil. Such processes are often called biomass-to-liquids or BTL.

Synthetic diesel may also be produced out of natural gas in the gas-to-liquid (GTL) process or out of coal in the coal-to-liquid (CTL) process. Such synthetic diesel has 30 percent lower particulate emissions than conventional diesel (US- California).

Biodiesel

Biodiesel can be obtained from vegetable oil (vegidiesel/vegifuel), or animal fats (bio-lipids), using transesterification. Biodiesel is a non-fossil fuel, cleaner burning alternative to petrodiesel. It can also be mixed with petrodiesel in any amount in some modern engines, but is 'strongly recommended against' by some manufacturers. Biodiesel has a higher gel point than petrodiesel, but is comparable to diesel. This can be overcome by using a biodiesel/petrodiesel blend, or by installing a fuel heater, but this is only necessary during the colder months. A small fraction of biodiesel can be used as an additive in low-sulfur formulations of diesel to increase the lubricity lost when the sulfur is removed. In the event of fuel spills, biodiesel is easily washed away with ordinary water and is nontoxic compared to other fuels.

Biodiesel made from soybean oil.

Biodiesel can be produced using kits. Certain kits allow for processing of used vegetable oil that can be run through any conventional diesel motor with modifications. The modification needed is the replacement of fuel lines from the intake and motor and all affected rubber fittings in injection and feeding pumps a.s.o (in vehicles manufactured before). This is because biodiesel is an effective solvent and will replace softeners within unsuitable rubber with itself over time. Synthetic gaskets for fittings and hoses prevent this.

Chemically, most biodiesel consists of alkyl (usually methyl) esters instead of the alkenes and aromatic hydrocarbons of petroleum derived diesel. However, biodiesel has combustion properties very similar to petrodiesel, including combustion energy and cetane ratings. Paraffin biodiesel also exists. Due to the purity of the source, it has a higher quality than petrodiesel.

Biodiesel Emissions

The use of biodiesel blended diesel fuels in fractions up to 99 percent result in substantial emission reductions. Sulfur oxide and sulfate emissions, major components of acid rain, are essentially eliminated with pure biodiesel and substantially reduced using biodiesel blends with minor quantities of ULSD petrodiesel. Use of biodiesel also results in substantial reductions of unburned hydrocarbons, carbon monoxide, and particulate matter compared to either gasoline or petrodiesel. CO, or carbon monoxide, emissions using biodiesel are substantially reduced, on the order of 50 percent compared to most petrodiesel fuels. The exhaust emissions of particulate matter from biodiesel have been found to be 30 percent lower than overall particulate matter emissions from petrodiesel. The exhaust emissions of total hydrocarbons (a contributing factor in the localized formation of smog and ozone) are up to 93 percent lower for biodiesel than diesel fuel. Biodiesel emissions of nitrogen oxides can sometimes increase slightly. However, biodiesel's complete lack of sulfur and sulfate emissions allows the use of NOx control technologies, such as AdBlue, that cannot be used with conventional diesel, allowing the management and control of nitrous oxide emissions.

Biodiesel also may reduce health risks associated with petroleum diesel. Biodiesel emissions showed decreased levels of PAH and nitrated PAH compounds which have been identified as potential cancer causing compounds. In recent testing, PAH compounds were reduced by 75 to 85 percent, with the exception of benzo(a)anthracene, which was reduced by roughly 50 percent.

Targeted nPAH compounds were also reduced dramatically with biodiesel fuel, with 2-nitrofluorene and 1-nitropyrene reduced by 90 percent, and the rest of the nPAH compounds reduced to only trace levels.

Transportation

Diesel fuel is widely used in most kinds of transportation. The gasoline-powered passenger automobile is the major exception.

Railroads

Diesel-electric locomotives are used predominantly on most railroads worldwide, except in areas such as a high percentage of the European continent where overhead electrification permits use of electric locomotives. Steam locomotives, dominant until the 1950s or 1960s in most regions, are now generally seen only on tourist-oriented historical railroads and special excursion trains.

Aircraft

The first diesel-powered flight of a fixed wing aircraft took place on the evening of September 18, 1928, at the Packard Motor Company proving grounds, Utica, Michigan with Captain Lionel M. Woolson and Walter Lees at the controls (the first "official" test flight was taken the next morning). The engine was designed for Packard by Woolson and the aircraft was a Stinson SM1B, X7654. Later that year Charles Lindbergh flew the same aircraft. In 1929 it was flown 621 miles (999 km) non-stop from Detroit to Langley, Virginia. This aircraft is presently owned by Greg Herrick and resides in the Golden Wings Flying Museum near Minneapolis, Minnesota. In 1931, Walter Lees and Fredrick Brossy set the nonstop flight record flying a Bellanca powered by a Packard diesel for 84h 32 m. The Hindenburg was powered by four 16-cylinder diesel engines, each with approximately 1,200 horsepower (890 kW) available in bursts, and 850 horsepower (630 kW) available for cruising. Modern diesel engines for propeller-driven aircraft are manufactured by Thielert Aircraft Engines and SMA. These engines are able to run on Jet A fuel, which is similar in composition to automotive diesel and cheaper and more plentiful than the 100 octane low-lead gasoline (avgas) used by the majority of the piston-engine aircraft fleet.

The most-produced aviation diesel engine in history so far has been the Junkers Jumo 205, which, along with its similar developments from the Junkers Motorenwerke, had approximately 1,000 examples of the unique opposed piston, two-stroke design powerplant built in the 1930s leading into World War II in Germany.

Automobiles

The very first diesel-engine automobile trip (inside the United States) was completed on January 6, 1930. The trip was from Indianapolis to New York City, a distance of nearly 800 miles (1,300 km). This feat helped to prove the usefulness of the compression ignition engine.

Automobile Racing

In 1931, Dave Evans drove his Cummins Diesel Special to a nonstop finish in the Indianapolis 500, the first time a car had completed the race without a pit stop. That car and a later Cummins Diesel Special are on display at the Indianapolis Motor Speedway Hall of Fame Museum.

In the late 1970s, Mercedes-Benz at Nardò drove a C111-III with a 5-cylinder diesel engine to several new records, including driving an average of 314 km/h (195 mph) for 12 hours and hitting a top speed of 325 km/h (201 mph). With turbocharged diesel cars getting stronger in the 1990s, they were entered in touring car racing, and BMW even won the 24 Hours with a 320d.

After winning the 12 Hours of Sebring in 2006 with the diesel-powered R10 TDI LMP, Audi won the 24 Hours of Le Mans, too. This is the first time a diesel-fueled vehicle has won at Le Mans against cars powered with regular fuel or other alternative fuel like methanol or bio-ethanol. French automaker Peugeot entered the diesel powered Peugeot 908 LMP in the 2007 24 Hours of Le Mans in response to the success of the Audi R10 TDI but Audi won the race again and for the third consecutive time in 2008. In 2008 Audi used next generation 10% BTL biodiesel manufactured from biomass.

In an effort to further demonstrate the potential of diesel power, California-based Gale Banks Engineering designed, built and raced a Cummins-powered pickup at the Bonneville Salt Flats in October 2002. The truck set a top speed of 355 km/h (222 mph) and became the world's fastest pickup, and almost equally notable, the truck drove to the race towing its own support trailer.

The British-based earthmoving machine manufacturer JCB raced the specially designed JCB Dieselmax car at 563.4 km/h (350.1 mph). The driver was Andy Green. The car was powered by two modified JCB 444 diesel engines.

Other important diesel engine performances are the SEAT León TDI's victories in the World Touring Car Championship.

Other uses

Poor quality (high sulfur) diesel fuel has been used as a palladium extraction agent for the liquid-liquid extraction of this metal from nitric acid mixtures. This has been proposed as a means of separating the fission product palladium from PUREX raffinate which comes from used nuclear fuel. In this solvent extraction system the hydrocarbons of the diesel act as the diluent while the dialkyl sulfides act as the extractant. This extraction operates by a solvation mechanism. So far neither a pilot plant nor full scale plant has been constructed to recover palladium, rhodium or ruthenium from nuclear wastes created by the use of nuclear fuel.

Health Effects

Diesel combustion exhaust is a major source of atmospheric soot and fine particles, constituting a portion of air pollutants implicated in human heart and lung damage. Diesel exhaust also contains nanoparticles. The study of nanoparticles and nanotoxicology is still in its infancy, and the full health effects from nanoparticles produced by all types of diesel are unknown. At least one study has observed that short-term exposure to diesel exhaust does not result in adverse extra-pulmonary effects, effects that are often correlated with an increase in cardiovascular disease. Long-term effects still need to be clarified, as well as the effects on susceptible groups of people with cardiopulmonary diseases.

Diesel exhaust from a large truck starting
up in USA with old technology device.

It should be noted that the types and quantities of nanoparticles can vary according to operating conditions, such as temperatures, pressures, presence of an open flame, fundamental fuel type and fuel mixture, and even atmospheric mixtures. As such, the types of nanoparticles produced by different engine technologies and even different fuels are not necessarily comparable. In general, the usage of biodiesel and biodiesel blends results in decreased pollution. One study has shown that the volatile component of 95 percent of diesel nanoparticles is unburned lubricating oil.

Taxation

Diesel fuel is very similar to heating oil used for central heating. In Europe, the United States, and Canada, taxes on diesel fuel are higher than on heating oil due to the fuel tax, and in those areas, heating oil is marked with fuel dyes and trace chemicals to prevent and detect tax fraud. Similarly, "untaxed" diesel (sometimes called "off road diesel") is available in the United States, which is available for use primarily in agricultural applications such as fuel for tractors, recreational and utility vehicles or other non-commercial vehicles that do not use public roads. Additionally, this fuel may have sulfur levels that exceed the limits for road use by the 2007 standards. This untaxed diesel is dyed red for identification purposes. Should a person be found using this untaxed diesel fuel for a typically taxed purpose (such as "over-the-road," or driving use), the user may have to pay a heavy fine.

In the United Kingdom, Belgium and the Netherlands it is known as red diesel (or gas oil), and is also used in agricultural vehicles, home heating tanks, refrigeration units on vans/trucks which contain perishable items (e.g. food, medicine) and for marine craft. Diesel fuel, or Marked Gas Oil is dyed green in the Republic of Ireland. The term DERV ("diesel engined road vehicle") is used in the UK as a synonym for unmarked road diesel fuel. In India, taxes on diesel fuel are lower than on petroleum as most vehicles that transport grain and other essential commodities across the country run on diesel.

In Germany, diesel fuel is taxed lower than petroleum but the annual vehicle tax is higher for diesel vehicles than for petroleum vehicles. This gives an advantage to vehicles that travel longer distances (which is the case for trucks and utility vehicles) because the annual vehicle tax depends only on engine displacement, not on distance driven. The point at which a diesel vehicle becomes less expensive than a comparable petroleum vehicle is around 20,000 km per year (12,500 miles per year) for an average car.

Taxes on biodiesel in the United States vary from state to state. Some states (Texas, for example) have no tax on biodiesel and a reduced tax on biodiesel blends. Other states, such as North Carolina, tax biodiesel (in any blended configuration) the same as petrodiesel, although they have introduced new incentives to producers and users of all biofuels.

Gasoline/Petrol

Gasoline (also called gas, petrol, or petrogasoline) is a petroleum-derived liquid mixture consisting mostly of aliphatic hydrocarbons, enhanced with iso-octane or the aromatic hydrocarbons toluene and benzene to increase its octane rating, and is primarily used as fuel in internal combustion engines.

Chemical Analysis and Production

Oil refineries produce gasoline.

Gasoline is produced in oil refineries. Material that is separated from crude oil via distillation, called virgin or straight-run gasoline, does not meet the required specifications for modern engines, but will form part of the blend.

The bulk of a typical gasoline consists of hydrocarbons with between 5 and 12 carbon atoms per molecule.

Many of these hydrocarbons are considered hazardous substances and are regulated in the United States by Occupational Safety and Health Administration. The Material Safety Data Sheet for unleaded gasoline shows at least fifteen hazardous chemicals occurring in various amounts. These include benzene (up to 5 percent by volume), toluene (up to 35 percent by volume), naphthalene (up to 1 percent by volume), trimethylbenzene (up to 7 percent by volume), MTBE (up to 18 percent by volume) and about ten others.

A United States nodding donkey.

An oil rig in the Gulf of Mexico.

The various refinery streams blended together to make gasoline all have different characteristics. Some important streams are:

- Reformate, produced in a catalytic reformer with a high octane rating and high aromatic content, and very low olefins (alkenes).

- Cat Cracked Gasoline or Cat Cracked Naphtha, produced from a catalytic cracker, with a moderate octane rating, high olefins (alkene) content, and moderate aromatics level. Here, "cat" is short for "catalytic".

- Hydrocrackate (Heavy, Mid, and Light), produced from a hydrocracker, with medium to low octane rating and moderate aromatic levels.

- Virgin or Straight-run Naphtha (has many names), directly from crude oil with low octane rating, low aromatics (depending on the crude oil), some naphthenes (cycloalkanes) and no olefins (alkenes).

- Alkylate, produced in an alkylation unit, with a high octane rating and which is pure paraffin (alkane), mainly branched chains.

- Isomerate (various names) which is obtained by isomerising the pentane and hexane in light virgin naphthas to yield their higher octane isomers.

Old petrol pumps in Nøtterøy, Norway.

The terms used here are not always the correct chemical terms. They are the jargon normally used in the oil industry. The exact terminology for these streams varies by refinery and by country.

- Overall a typical gasoline is predominantly a mixture of paraffins (alkanes), naphthenes (cycloalkanes), and olefins (alkenes). The exact ratios can depend on The oil refinery that makes the gasoline, as not all refineries have the same set of processing units.

- The crude oil feed used by the refinery.

- The grade of gasoline, in particular the octane rating.

Currently many countries set tight limits on gasoline aromatics in general, benzene in particular, and olefin (alkene) content. This is increasing the demand for high octane pure paraffin (alkane) components, such as alkylate, and is forcing refineries to add processing units to reduce the benzene content.

Gasoline can also contain some other organic compounds: Such as organic ethers (deliberately added), plus small levels of contaminants, in particular sulfur compounds such as disulfides and thiophenes. Some contaminants, in particular thiols and hydrogen sulfide, must be removed because they cause corrosion in engines. Sulfur compounds are usually removed by hydrotreating, yielding hydrogen sulfide which can then be transformed into elemental sulfur via the Claus process.

The density of gasoline is 0.71–0.77 g/cm³, (in English units, approx. 0.026 lb/cu in or 6.073 lb/U.S. gal or 7.29 lb/imp gal) which means it floats on water. This may be advantageous in the event of a spill. It is flammable and can burn while floating over water.

Volatility

A container for storing gasoline used in Germany.

Gasoline is more volatile than diesel oil, Jet-A or kerosene, not only because of the base constituents, but because of the additives that are put into it. The final control of volatility is often achieved by blending with butane. The Reid Vapor Pressure test is used to measure the volatility of gasoline. The desired volatility depends on the ambient temperature: in hotter climates, gasoline components of higher molecular weight and thus lower volatility are used. In cold climates, too little volatility results in cars failing to start. In hot climates, excessive volatility results in what is known as "vapor lock" where combustion fails to occur, because the liquid fuel has changed to a gaseous fuel in the fuel lines, rendering the fuel pump ineffective and starving the engine of fuel.

In the United States, volatility is regulated in large urban centers to reduce the emission of unburned hydrocarbons. In large cities, so-called reformulated gasoline that is less prone to evaporation, among other properties, is required. In Australia, summer petrol volatility limits are set by State Governments and vary between capital cities. Most countries simply have a summer, winter, and perhaps intermediate limit.

Volatility standards may be relaxed (allowing more gasoline components into the atmosphere) during emergency anticipated gasoline shortages. For example, on August 31, 2005, in response to Hurricane Katrina, the United States permitted the sale of non-reformulated gasoline in some urban areas, which effectively permitted an early switch from summer to winter-grade gasoline. As mandated by EPA administrator Stephen L. Johnson, this "fuel waiver" was made effective through September 15, 2005. Though relaxed volatility standards may increase the atmospheric concentration of volatile organic compounds in warm weather, higher volatility gasoline effectively increases a nation's gasoline supply because the amount of butane in the gasoline pool is allowed to increase.

Octane Rating

An important characteristic of gasoline is its octane rating, which is a measure of how resistant gasoline is to the abnormal combustion phenomenon known as detonation (also known as knocking, pinking, spark knock, and other names). Deflagration is the normal type of combustion. Octane rating is measured relative to a mixture of 2,2,4-trimethylpentane (an isomer of octane) and n-heptane. There are a number of different conventions for expressing the octane rating; therefore, the same fuel may be labeled with a different number, depending upon the system used.

The octane rating became important in the search for higher output powers from aero engines in the late 1930s and the 1940s as it allowed higher compression ratios to be used.

Energy Content

A container for storing gasoline used in the
United States; red containers are typically used.

Gasoline contains about 34.8 MJ/L or 132 MJ/US gallon. This is about 9.67 kWh/L or 36.6 kWh/U.S. gallon. This is an average; gasoline blends differ, therefore actual energy content varies from season to season and from batch to batch, by up to 4 percent more or less than the average, according to the U.S. EPA. On average, about 19.5 US gallons (16.2 imp gal/74 L) of gasoline are available from a Template:Convert/LoffAoffDbSonUSre barrel of crude oil (about 46 percent by volume), varying due to quality of crude and grade of gasoline. The remaining residue comes off as products ranging from tar to naptha.

Table: Volumetric energy density of some fuels compared with gasoline.

Fuel type	MJ/litre	MJ/kg	BTU/Imp gal	BTU/US gal	Research octane number (RON)
Regular Gasoline	34.8	44.4	150,100	125,000	Min 91
Premium Gasoline	39.5				Min 95
Autogas (LPG) (60% Propane + 40% Butane)	26.8	46			108
Ethanol	23.5	31.1	101,600	84,600	129
Methanol	17.9	19.9	77,600	64,600	123
Butanol	29.2				91-99
Gasohol (10% ethanol + 90% gasoline)	33.7		145,200	120,900	93/94
Diesel	38.6	45.4	166,600	138,700	25(*)
Aviation gasoline (high octane gasoline, not jet fuel)	33.5	46.8	144,400	120,200	
Jet fuel (kerosene based)	35.1	43.8	151,242	125,935	
Liquefied natural gas	25.3	~55	109,000	90,800	
Hydrogen		121			130

(*) Diesel is not used in a gasoline engine, so its low octane rating is not an issue; the relevant metric for diesel engines is the cetane number.

A high octane fuel such as Liquefied petroleum gas (LPG) has a lower energy content than lower octane gasoline, resulting in an overall lower power output at the regular compression ratio an engine ran at on gasoline. However, with an engine tuned to the use of LPG (that is, via higher compression ratios such as 12:1 instead of 8:1), this lower power output can be overcome. This is because higher-octane fuels allow for a higher compression ratio—this means less space in a cylinder on its combustion stroke, hence a higher cylinder temperature which improves efficiency according to Carnot's theorem, along with fewer wasted hydrocarbons (therefore less pollution and wasted energy), bringing higher power levels coupled with less pollution overall because of the greater efficiency.

The main reason for the lower energy content (per litre) of LPG in comparison to gasoline is that it has a lower density. Energy content per kilogram is higher than for gasoline (higher hydrogen to carbon ratio). The weight-density of gasoline is about 740 kg/m³ (6.175 lb/US gal; 7.416 lb/imp gal).

Different countries have some variation in what RON (Research Octane Number) is standard for gasoline, or petrol. In the UK, ordinary regular unleaded petrol is 91 RON (not commonly available), premium unleaded petrol is always 95 RON, and super unleaded is usually 97-98 RON. However both Shell and BP produce fuel at 102 RON for cars with hi-performance engines, and the supermarket chain Tesco began in 2006 to sell super unleaded petrol rated at 99 RON. In the U.S., octane ratings in unleaded fuels can vary between 86-87 AKI (91-92 RON) for regular, through

89-90 AKI (94-95 RON) for mid-grade (European Premium), up to 90-94 AKI (95-99 RON) for premium (European Super).

Additives

Lead

The mixture known as gasoline, when used in high compression internal combustion engines, has a tendency to autoignite*(detonation)* causing a damaging "engine knocking" (also called "pinging") noise. Early research into this effect was led by A.H. Gibson and Harry Ricardo in England and Thomas Midgley and Thomas Boyd in the United States. The discovery that lead additives modified this behavior led to the widespread adoption of the practice in the 1920s and therefore more powerful higher compression engines. The most popular additive was tetra-ethyl lead. However, with the discovery of the environmental and health damage caused by the lead, and the incompatibility of lead with catalytic converters found on virtually all newly sold U.S. automobiles since 1975, this practice began to wane (encouraged by many governments introducing differential tax rates) in the 1980s. Most countries are phasing out leaded fuel; different additives have replaced the lead compounds. The most popular additives include aromatic hydrocarbons, ethers and alcohol (usually ethanol or methanol).

In the U.S., where lead had been blended with gasoline (primarily to boost octane levels) since the early 1920s, standards to phase out leaded gasoline were first implemented in 1973. In 1995, leaded fuel accounted for only 0.6 percent of total gasoline sales and less than 2,000 short tons of lead per year. From January 1, 1996, the Clean Air Act banned the sale of leaded fuel for use in on-road vehicles. Possession and use of leaded gasoline in a regular on-road vehicle now carries a maximum $10,000 fine in the US. However, fuel containing lead may continue to be sold for off-road uses, including aircraft, racing cars, farm equipment, and marine engines. The ban on leaded gasoline led to thousands of tons of lead not being released in the air by automobiles. Similar bans in other countries have resulted in lowering levels of lead in people's bloodstreams.

A side effect of the lead additives was protection of the valve seats from erosion. Many classic cars' engines have needed modification to use lead-free fuels since leaded fuels became unavailable. However, "Lead substitute" products are also produced and can sometimes be found at auto parts stores. These were scientifically tested and some were approved by the Federation of British Historic Vehicle Clubs at the UK's Motor Industry Research Association (MIRA) in 1999.

Gasoline, as delivered at the pump, also contains additives to reduce internal engine carbon build-ups, improve combustion, and to allow easier starting in cold climates.

In some parts of South America, Asia, Eastern Europe and the Middle East, leaded gasoline is still in use. Leaded gasoline was phased out in sub-Saharan Africa effective January 1, 2006. A growing number of countries have drawn up plans to ban leaded gasoline in the near future.

MMT

Methylcyclopentadienyl manganese tricarbonyl (MMT) has been used for many years in Canada and recently in Australia to boost octane. It also helps old cars designed for leaded fuel run on unleaded fuel without need for additives to prevent valve problems.

U.S. Federal sources state that MMT is suspected to be a powerful neurotoxin and respiratory toxin, and a large Canadian study concluded that MMT impairs the effectiveness of automobile emission controls and increases pollution from motor vehicles.

In 1977, use of MMT was banned in the U.S. by the Clean Air Act until the Ethyl Corporation could prove that the additive would not lead to failure of new car emissions-control systems. As a result of this ruling, the Ethyl Corporation began a legal battle with the EPA, presenting evidence that MMT was harmless to automobile emissions-control systems. In 1995, the U.S. Court of Appeals ruled that the EPA had exceeded its authority, and MMT became a legal fuel additive in the U.S. MMT is nowadays manufactured by the Afton Chemical Corporation division of Newmarket Corporation.

Ethanol

In the United States, ethanol is sometimes added to gasoline but sold without an indication that it is a component. Chevron, 76, Shell, and several other brands market ethanol-gasoline blends.

In several states, ethanol is added by law to a minimum level which is currently 5.9 percent. Most fuel pumps display a sticker stating that the fuel may contain up to 10 percent ethanol, an intentional disparity which allows the minimum level to be raised over time without requiring modification of the literature/labeling. The bill which was being debated at the time the disclosure of the presence of ethanol in the fuel was mandated has recently passed. This law (Energy Policy Act of 2005) will require all auto fuel to contain at least 10 percent ethanol. Many call this fuel mix gasohol.

In the EU, 5 percent ethanol can be added within the common gasoline spec (EN 228). Discussions are ongoing to allow 10 percent blending of ethanol. Most countries (fuel distributors) today do not add so much ethanol. Most gasoline (petrol) sold in Sweden has 5 percent ethanol added.

In Brazil, the Brazilian National Agency of Petroleum, Natural Gas and Biofuels (ANP) requires that gasoline for automobile use has 23 percent of ethanol added to its composition.

Dye

In the United States the most commonly used aircraft gasoline, avgas, or aviation gas, is known as 100LL (100 octane, low lead) and is dyed blue. Red dye has been used for identifying untaxed (non-highway use) agricultural diesel. The UK uses red dye to differentiate between regular diesel fuel, (often referred to as DERV from *Diesel-Engined Road Vehicle*), which is undyed, and diesel intended for agricultural and construction vehicles like excavators and bulldozers. Red diesel is still occasionally used on HGVs which use a separate engine to power a loader crane. This is a declining practice however, as many loader cranes are powered directly by the tractor unit.

Oxygenate Blending

Oxygenate blending adds oxygen to the fuel in oxygen-bearing compounds such as MTBE, ETBE, and ethanol, and so reduces the amount of carbon monoxide and unburned fuel in the exhaust gas, thus reducing smog. In many areas throughout the U.S. oxygenate blending is mandated by EPA regulations to reduce smog and other airborne pollutants. For example, in Southern California, fuel must contain 2 percent oxygen by weight, resulting in a mixture of 5.6 percent ethanol in gasoline. The resulting fuel is often known as *reformulated gasoline* (RFG) or *oxygenated gasoline*. The federal requirement that RFG contain oxygen was dropped May 6, 2006, because the industry had developed VOC-controlled RFG that did not need additional oxygen.

MTBE use is being phased out in some states due to issues with contamination of ground water. In some places, such as California, it is already banned. Ethanol and to a lesser extent the ethanol derived ETBE are a common replacements. Especially since ethanol derived from biomatter such as corn, sugar cane or grain is frequent, this will often be referred to as *bio*-ethanol. A common ethanol-gasoline mix of 10 percent ethanol mixed with gasoline is called gasohol or E10, and an ethanol-gasoline mix of 85% ethanol mixed with gasoline is called E85. The most extensive use of ethanol takes place in Brazil, where the ethanol is derived from sugarcane. In 2004, over 3.4 billion U.S. gallons (2.8 billion imp gal/13 million m³) of ethanol was produced in the United States for fuel use, mostly from corn, and E85 is slowly becoming available in much of the United States. Unfortunately many of the relatively few stations vending E85 are not open to the general public. The use of bioethanol, either directly or indirectly by conversion of such ethanol to *bio*-ETBE, is encouraged by the European Union Directive on the Promotion of the use of biofuels and other renewable fuels for transport. However since producing bio-ethanol from fermented sugars and starches involves distillation, ordinary people in much of Europe cannot legally ferment and distill their own bio-ethanol at present (unlike in the U.S. where getting a BATF distillation permit has been easy since the 1973 oil crisis).

Health Concerns

Uncontrolled burning of gasoline
produces large quantities of soot.

Many of the non-aliphatic hydrocarbons naturally present in gasoline (especially aromatic ones like benzene), as well as many anti-knocking additives, are carcinogenic. Because of this, any large-scale or ongoing leaks of gasoline pose a threat to the public's health and the environment, should the gasoline reach a public supply of drinking water. The chief risks of such leaks come not from vehicles, but from gasoline delivery truck accidents and leaks from storage tanks. Because of this risk, most (underground) storage tanks now have extensive measures in place to detect and prevent any such leaks, such as sacrificial anodes. Gasoline is rather volatile (meaning it readily evaporates), requiring that storage tanks on land and in vehicles be properly sealed. The high volatility also means that it will easily ignite in cold weather conditions, unlike diesel for example. Appropriate venting is needed to ensure the level of pressure is similar on the inside and outside. Gasoline also reacts dangerously with certain common chemicals.

Gasoline is also one of the sources of pollutant gases. Even gasoline which does not contain lead or sulfur compounds produces carbon dioxide, nitrogen oxides, and carbon monoxide in the exhaust of the engine which is running on it. Furthermore, unburnt gasoline and evaporation from the tank, when in the atmosphere, react in sunlight to produce photochemical smog. Addition of ethanol increases the volatility of gasoline.

Through misuse as an inhalant, gasoline also contributes to damage to health. Petrol sniffing is a common way of obtaining a high for many people and has become epidemic in some poorer communities and indigenous groups in America, Australia, Canada, New Zealand and some Pacific Islands. In response, Opal fuel has been developed by the BP Kwinana Refinery in Australia, and contains only 5 percent aromatics (unlike the usual 25 percent) which inhibits the effects of inhalation.

Like other alkenes, gasoline burns in the vapor phase and, coupled with its volatility, this makes leaks highly dangerous when sources of ignition are present. Many accidents involve gasoline being used in an attempt to light bonfires; rather than helping the material on the bonfire to burn, some of the gasoline vaporizes quickly after being poured and mixes with the surrounding air, so when the fire is lit a moment later the vapor surrounding the bonfire instantly ignites in a large fireball, engulfing the unwary user. The vapor is also heavier than air and tends to collect in garage inspection pits.

Usage and Pricing

UK petrol prices.

The U.S. accounts for about 44 percent of the world's gasoline consumption. In 2003, The U.S. consumed Template:Convert/GL, which equates to 1.3 gigalitres of gasoline each day (about 360 million U.S. gallons or 300 million imperial gallons). The U.S. used about 510 billion liters (138 billion U.S. gal/115 billion imp gal) of gasoline in 2006, of which 5.6 percent was mid-grade and 9.5 percent was premium grade. Western countries have among the highest usage rates per person.

Based on externalities, some countries, for example, in Europe and Japan, impose heavy fuel taxes on fuels such as gasoline. Because a greater proportion of the price of gasoline in the United States is due to the cost of oil, rather than taxes, the price of the retail product is subject to greater fluctuations (vs. outside the U.S.) when calculated as a percentage of cost-per-unit, but is actually less variable in absolute terms.

Stability

When gasoline is left for a period of time, gums and varnishes may build up and precipitate in the gasoline, causing "stale fuel." This will cause gums to build up in the fuel tank, lines, and carburetor or fuel injection components making it harder to start the engine. Motor gasoline may be stored up to 60 days in an approved container. If it is to be stored for a longer period of time, a fuel stabilizer may be used. This will extend the life of the fuel to about 1-2 years, and keep it fresh for the next uses. Fuel stabilizer is commonly used for small engines such as lawnmower and tractor engines to promote quicker and more reliable starting. Users have been advised to keep gasoline containers and tanks more than half full and properly capped to reduce air exposure, to avoid storage at high temperatures, to run an engine for ten minutes to circulate the stabilizer through all components prior to storage, and to run the engine at intervals to purge stale fuel from the carburetor.

Gummy, sticky resin deposits result from oxidative degradation of gasoline. This degradation can be prevented through the use of antioxidants such as phenylenediamines, alkylenediamines (diethylenetriamine, triethylenetetramine, etc), and alkylamines (diethylamine, tributylamine, ethylamine). Other useful additives include gum inhibitors such as N-substituted alkylaminophenols and color stabilizers such as N-(2-aminoethyl)piperazine, N,N-diethylhydroxylamine, and triethylenetetramine.

By 1975, improvements in refinery techniques have generally reduced the reliance on the catalytically or thermally cracked stocks most susceptible to oxidation. Gasoline containing acidic contaminants such as naphthenic acids can be addressed with additives including strongly basic organo-amines such as N,N-diethylhydroxylamine, preventing metal corrosion and breakdown of other antioxidant additives due to acidity. Hydrocarbons with a bromine number of 10 or above can be protected with the combination of unhindered or partially hindered phenols and oil soluble strong amine bases such as monoethanolamine, N-(2-aminoethyl)piperazine, cyclohexylamine, 1,3-cyclohexane-bis(methylamine), 2,5-dimethylaniline, 2,6-dimethylaniline, diethylenetriamine, and triethylenetetramine.

"Stale" gasoline can be detected by a colorimetric enzymatic test for organic peroxides produced by oxidation of the gasoline.

Other Fuels

Many of these alternatives are less damaging to the environment than gasoline, but the first generation biofuels are still not 100 percent clean.

- Biofuels:
  - Biodiesel, for diesel engines.

- ○ Biobutanol, for gasoline engines.

- ○ Bioethanol.

- ○ Biogasoline.

- Compressed air.

- Hydrogen fuel.

- Electricity.

- Fossil fuels:

  - ○ CNG (Compressed Natural Gas).

  - ○ Petrodiesel.

Bioconversion and Biogasoline

XcelPlus Global Holdings, working in conjunction with Maverick BioFuels, developed the technology in which a fuel compatible with internal combustion gasoline engines is derived from natural renewable oils like soybean, other vegetable oils and biodiesel. Initial marketing efforts will focus on an additive package for converting ordinary Biodiesel into gasoline, adding the Biolene additive package. The additive is expected to be on the market later this year. Home blenders can expect final pump-grade fuel to cost approximately US$2.70 per U.S. gallon ($3.24/imp gal, 71¢/L).

Companies such as Sapphire Energy are developing a means to "grow" gasoline, that is, produce it directly from living organisms (that is, algae). Biogasoline has the advantage of not needing any change in vehicle or distribution infrastructure.

Paraffin Wax

Paraffin wax is colourless or white somewhat translucent, hard wax consisting of a mixture of solid straight-chain hydrocarbons ranging in melting point from about 48° to 66 °C (120° to 150 °F). Paraffin wax is obtained from petroleum by dewaxing light lubricating oil stocks. It is used in candles, wax paper, polishes, cosmetics, and electrical insulators. It assists in extracting perfumes from flowers, forms a base for medical ointments, and supplies a waterproof coating for wood. In wood and paper matches, it helps to ignite the matchstick by supplying an easily vaporized hydrocarbon fuel.

Paraffin wax was first produced commercially in 1867, less than 10 years after the first petroleum well was drilled. Paraffin wax precipitates readily from petroleum on chilling. Technical progress has served only to make the separations and filtration more efficient and economical. Purification methods consist of chemical treatment, decolorization by adsorbents, and fractionation of the separated waxes into grades by distillation, recrystallization, or both. Crude oils differ widely in wax content.

Synthetic paraffin wax was introduced commercially after World War II as one of the products obtained in the Fischer–Tropsch reaction, which converts coal gas to hydrocarbons. Snow-white and harder than petroleum paraffin wax, the synthetic product has a unique character and high purity that make it a suitable replacement for certain vegetable waxes and as a modifier for petroleum waxes and for some plastics, such as polyethylene. Synthetic paraffin waxes may be oxidized to yield pale-yellow, hard waxes of high molecular weight that can be saponified with aqueous solutions of organic or inorganic alkalies, such as borax, sodium hydroxide, triethanolamine, and morpholine. These wax dispersions serve as heavy-duty floor wax, as waterproofing for textiles and paper, as tanning agents for leather, as metal-drawing lubricants, as rust preventives, and for masonry and concrete treatment.

Manufacturing

The feedstock for paraffin is slack wax, which is a mixture of oil and wax, a byproduct from the refining of lubricating oil.

The first step in making paraffin wax is to remove the oil (de-oiling or de-waxing) from the slack wax. The oil is separated by crystallization. Most commonly, the slack wax is heated, mixed with one or more solvents such as a ketone and then cooled. As it cools, wax crystallizes out of the solution, leaving only oil. This mixture is filtered into two streams: solid (wax plus some solvent) and liquid (oil and solvent). After the solvent is recovered by distillation, the resulting products are called "product wax" (or "press wax") and "foots oil". The lower the percentage of oil in the wax, the more refined it is considered (semi-refined versus fully refined). The product wax may be further processed to remove colors and odors. The wax may finally be blended together to give certain desired properties such as melt point and penetration. Paraffin wax is sold in either liquid or solid form.

Applications

In industrial applications, it is often useful to modify the crystal properties of the paraffin wax, typically by adding branching to the existing carbon backbone chain. The modification is usually done with additives, such as EVA copolymers, microcrystalline wax, or forms of polyethylene. The branched properties result in a modified paraffin with a higher viscosity, smaller crystalline structure, and modified functional properties. Pure paraffin wax is rarely used for carving original models for casting metal and other materials in the lost wax process, as it is relatively brittle at room temperature and presents the risks of chipping and breakage when worked. Soft and pliable waxes, like beeswax, may be preferred for such sculpture, but "investment casting waxes," often paraffin-based, are expressly formulated for the purpose.

In a pathology laboratory, paraffin wax is used to impregnate tissue prior to sectioning thin samples of tissue. Water is removed from the tissue through ascending strengths of alcohol (75% to absolute) and the tissue is cleared in an organic solvent such as xylene. The tissue is then placed in paraffin wax for a number of hours and then set in a mold with wax to cool and solidify; sections are then cut on a microtome.

Other uses

- Candle-making.
- Wax carving.

- Coatings for waxed paper or cloth.

- Food-grade paraffin wax:

  ○ Shiny coating used in candy-making; although edible, it is nondigestible, passing through the body without being broken down.

  ○ Coating for many kinds of hard cheese, like Edam cheese.

  ○ Sealant for jars, cans, and bottles.

  ○ Chewing gum additive.

- Investment casting.

- Anti-caking agent, moisture repellent, and dustbinding coatings for fertilizers.

- Agent for preparation of specimens for histology.

- Bullet lubricant – with other ingredients, such as olive oil and beeswax.

- Phlegmatizing agent, commonly used to stabilise/desensitize high explosives such as RDX.

- Crayons.

- Solid propellant for hybrid rocket motors.

- Component of surfwax, used for grip on surfboards in surfing.

- Component of glide wax, used on skis and snowboards.

- Friction-reducer, for use on handrails and cement ledges, commonly used in skateboarding.

- Ink. Used as the basis for solid ink different color blocks of wax for thermal printers. The wax is melted and then sprayed on the paper producing images with a shiny surface.

- Microwax: food additive, a glazing agent with E number E905.

- Forensic investigations: the nitrate test uses paraffin wax to detect nitrates and nitrites on the hand of a shooting suspect.

- Antiozonant agents: blends of paraffin and micro waxes are used in rubber compounds to prevent cracking of the rubber; the admixture of wax migrates to the surface of the product and forms a protective layer. The layer can also act as a release agent, helping the product separate from its mould.

- Mechanical thermostats and actuators, as an expansion medium for activating such devices.

- "Potting" guitar pickups, which reduces microphonic feedback caused from the subtle movements of the pole pieces.

- "Potting" of local oscillator coils to prevent microphonic frequency modulation in low end FM radios.

- Textile manufacturing processes, such as that used for Eisengarn thread.

- Wax baths for beauty and therapy purposes.

- Thickening agent in many paintballs.

- Moisturiser in toiletries and cosmetics such as Vaseline.

- Prevents oxidation on the surface of polished steel and iron.

- Phase change material for thermal energy storage:

    ◦ MESSENGER (Mercury spacecraft) When the spacecraft was unable to radiate excessive heat.

- Manufacture of boiled leather armor and books.

- Skateboard wax.

- Paraffin microactuator.

- Neutron radiation shielding.

- Waterproofing agent for waxed cotton garments and commercially important in the early water proofing of ship sails.

- In Occupational and Physical therapies paraffin wax baths are used to warm and loosen connective tissue. They are mainly used in hand therapy.

- Used as the main additive in Log Sealers.

Occupational Safety

People can be exposed to paraffin in the workplace by breathing it in, skin contact, and eye contact. The National Institute for Occupational Safety and Health (NIOSH) has set a recommended exposure limit (REL) for paraffin wax fume exposure of 2 mg/m^3 over an 8-hour workday.

Kerosene

Kerosene is a flammable liquid mixture of chemicals that are produced in the distillation of crude oil. To produce kerosene, crude oil is distilled in a distillation tower in a process similar to that used to produce diesel and gasoline. It is a medium weight distillate in the refining process, and can be produced by distilling crude oil (here it is known as straight run kerosene) or by hydrocarbon cracking heavier petroleum (here it is known as cracked kerosene). The chemical composition of kerosene is fairly complex, and it is a complex mixture of paraffins (55.2%), naphthenes (40.9%), and aromatic hydrocarbons (3.9%). Kerosene tends to contain hydrocarbons that have anywhere from 11 to 13 carbons in the chains. Liquid kerosene fuels contain potentially harmful compounds, including hexane and benzene.

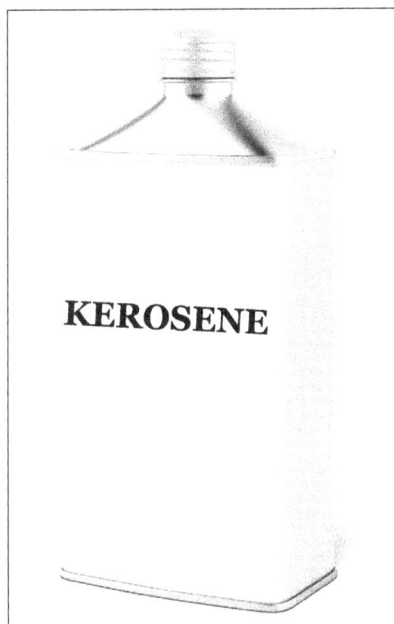

A bottle of blue-dyed kerosene.

Health Risks

The use of kerosene as an oil in heaters can be dangerous and because of that it is not used frequently. When operating, kerosene heaters can cause degradation of air quality inside a home while producing toxic and carcinogenic gases. Because of this, kerosene is not actively used in home heating in most developed countries.

In developing countries, the widespread use of kerosene comes with numerous different issues. Hazards of kerosene use include poisoning, fires, and explosions. As well, some kerosene lamps emit fine particulates, carbon monoxide, nitric oxides (NOx), and sulfur dioxide when burned. These by-products may reduce lung function and increase risks of asthma and cancer. Taking into account the risks of using kerosene, cleaner alternatives to kerosene technologies for lighting and cooking should be investigated - although kerosene is still a safer option in many cases than using solid fuels.

Climate Issues

Generally speaking, kerosene lamps are inefficient and produce harmful by-products of combustion when used. When kerosene is burned in wick lamps, about 7-9% of the kerosene consumed is converted to particulate matter that is almost entirely black carbon - a harmful emission. When wood is burned, less than half a percent is turned to black carbon. Reduction of black carbon emissions is marked as a potential way to reduce climate warming, and since kerosene lamps are such a major sources of black carbon, limiting their use could be beneficial to the environment. Alternatives to this kerosene use involve more electrification or using cheap LED lamps.

Uses

Kerosene is a major component of aviation fuel, making up more than 60% of the fuel. In addition, it can be used as an oil in central heating systems and can be used as a cleaning agent.

Although the use of kerosene in many places has decreased over the years as a result of improved access to electricity and natural gas, it is still used extensively in the developing world for cooking, heating, and lighting. Kerosene cooking is extensive in developing countries, especially among urban populations. Kerosene is often seen as a good alternative to solid fuels, biomass, and coal and thus kerosene lanterns are used in places where access to electricity is not available. It is estimated that globally 500 million households use fuels such as kerosene for lighting.

Kerosene oil has been used since the mid-19th century, when it replaced whale oil as a lighting fuel. Even though its use for lighting has been replaced by electricity in most parts of the world, kerosene oil lamps are still in use. There are other terms for kerosene oil, including paraffin, fuel oil number one, lamp oil and paraffin oil. Originally, kerosene oil was produced from coal and was often referred to as coal oil. Today, it is made from petroleum oil through fractional distillation.

While kerosene is not used as much for heating, it is still used in portable kerosene room heaters. It is also used for some appliances, such as kerosene stoves. Although rare, there are also kerosene refrigerators and other appliances that use kerosene to operate.

Care should always be taken when using kerosene for heating, cooking or lighting. It does have a low vapor pressure (high flash point) compared to gasoline or LPG, and thus has a lower risk of explosion. However, care should be taken to avoid a fire with any open flame, as spillage can result in a rapidly spreading fire. It is wise to keep loose-fitting, flammable clothing away from a kerosene oil lamp or cooking apparatus.

Probably one of the most common uses of kerosene oil is as lamp oil. Households living in areas where there is danger of power outages should have a few oil lamps and a supply of kerosene oil on hand, taking care to store them properly. People in areas that experience cold winter temperatures and rely on electricity for heat should also consider having a kerosene heater available as a backup.

Petroleum Coke

Petroleum coke is a byproduct of the oil refining process. As refineries worldwide seek to operate more efficiently and extract more gasoline and other high value fuels from each barrel of crude oil, a solid carbon material known as petcoke is produced.

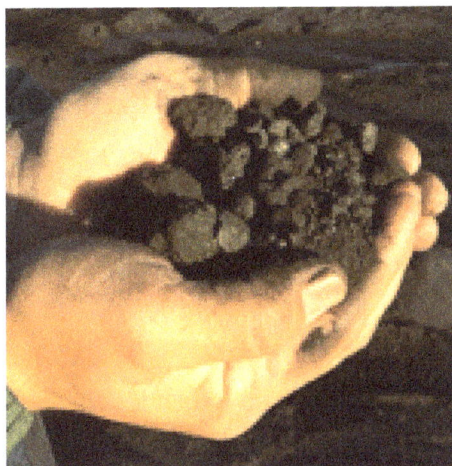

The chemical and physical characteristics of petcoke are a function of the crude oil and refining technology used by the refinery. Petcoke can be hard or relatively soft. Physically, petcoke can resemble large sponges with numerous pores, or it can resemble small spheres, ranging in size from a grain of sand to a large marble. Chemically, petcoke can include a variety of elements and metals in a wide range of concentrations. Depending on these physical and chemical characteristics, petcoke is typically used in either an energy application, as a source of British Thermal Units (BTUs) or in an industrial application, as a source of carbon.

Fuel grade petcoke represents roughly 80 percent of worldwide petcoke production, and Oxbow is the worldwide leader in fuel grade petcoke sourcing and sales, handling more than 11 million tons per year. Oxbow sources petcoke from every major refining company in the world, including ExxonMobil, Valero, Chevron, British Petroleum, PBF Energy, Phillips 66, Tesoro, Essar, Reliance and Shell.

Fuel grade petcoke is typically very high in heating value (BTUs per pound), produces virtually no ash when burned, and is most commonly used in electric power plants and cement kilns.

Types

There are at least four basic types of petroleum coke, namely, needle coke, honeycomb coke, sponge coke and shot coke. Different types of petroleum coke have different microstructures due to differences in operating variables and nature of feedstock. Significant differences are also to be observed in the properties of the different types of coke, particularly ash and volatile matter contents.

Needle coke, also called acicular coke, is a highly crystalline petroleum coke used in the production of electrodes for the steel and aluminium industries and is particularly valuable because the electrodes must be replaced regularly. Needle coke is produced exclusively from either FCC decant oil or coal tar pitch.

Honeycomb coke is an intermediate coke, with ellipsoidal pores that are uniformly distributed. Compared to needle coke, honeycomb coke has a lower coefficient of thermal expansion and a lower electrical conductivity.

Composition

Petcoke, altered through the process of calcined which it is heated or refined raw coke eliminates much of the component of the resource. Usually petcoke when refined does not release the heavy metals as volatiles or emissions.

Depending on the petroleum feed stock used, the composition of petcoke may vary but the main thing is that it is primarily carbon. Petcoke is primarily made up of carbon, when in pure form petcoke can weigh 98-99% which creates a carbon based compound with the hydrogen filling in. In raw form hydrogen can have a weight range of 3.0- 4.0%. Petcoke in its raw(green coke) nitrogen at 0.1- 0.5% and sulfur 0.2- 6.0% become emissions after the coke calcined.

Composition of raw petcoke	
Component	Raw (green) coke
Carbon (wt%)	80-95
Hydrogen (wt%)	3.0-4.5
Nitrogen (wt%)	0.1- 0.5
Sulfur (wt%)	0.2-6.0
Volatile matter (wt%)	5.0-15
Moisture (wt%)	0.5-10
Ash (wt%)	0.1- 1.0
Density (wt%)	1.2- 1.6
Heavy Metals (ppm. wt)	
Aluminium	15-100
Boron	0.1- 15
Calcium	25- 500
Chromium	5-50
Cobalt	10-60
Iron	50-5000
Manganese	2-100
Magnesium	10-250
Molybdenum	10-20
Nickel	10-500
Potassium	20-50
Silicon	50-600
Sodium	40-70
Titanium	2-60
Vanadium	5-500

Through process of thermal processing the composition in weight is reduced with the volatile matter and sulfur being emitted. This process ends in the honeycomb petcoke which according to the name giving is a solid carbon structure with holes in it.

Component	Petcoke (Calcined @ 2375 °F)
Carbon (wt%)	98.0-99.5
Hydrogen (wt%)	0.1
Nitrogen (wt%)	
Sulfur (wt%)	
Volatile matter (wt%)	0.2-0.8
Moisture (wt%)	0.1
Ash (wt%)	0.02-0.7
Density (wt%)	1.9-2.1
Heavy Metals (ppm. wt)	
Aluminium	15-100
Boron	0.1- 15
Calcium	25- 500
Chromium	5-50
Cobalt	10-60
Iron	50-5000
Manganese	2-100
Magnesium	10-250
Molybdenum	10-20
Nickel	10-500
Potassium	20-50
Silicon	50-600
Sodium	40-70
Titanium	2-60
Vanadium	5-500

Depending on the petroleum feed stock used, the composition of petcoke may vary but the main thing is that it is primarily carbon. Petcoke is primarily made up of carbon, when in pure form petcoke can weigh 98-99% which creates a carbon based compound with the hydrogen filling in. In raw form hydrogen can have a weight range of 3.0- 4.0%. Petcoke in its raw(green coke) nitrogen at 0.1- 0.5% and sulfur 0.2- 6.0% become emissions after the coke calcined.

Other heavy metals found can be found with in petcoke as impurities due to that some of these metals come in after processing as volatile.

Fuel-grade

Fuel-grade coke is classified as either sponge coke or shot coke morphology. While oil refiners have been producing coke for over 100 years, the mechanisms that cause sponge coke or shot coke to form are not well understood and cannot be accurately predicted. In general, lower temperatures and higher pressures promote sponge coke formation. Additionally, the amount of heptane insolubles present and the fraction of light components in the coker feed contribute.

While its high heat and low ash content make it a decent fuel for power generation in coal-fired boilers, petroleum coke is high in sulfur and low in volatile content, and this poses environmental (and technical) problems with its combustion. Its gross calorific value (HHV) is nearly 8000 Kcal/ kg which is twice the value of average coal used in electricity generation. A common choice of sulfur recovering unit for burning petroleum coke is the SNOX Flue gas desulfurisation technology, which is based on the well-known WSA Process. Fluidized bed combustion is commonly used to burn petroleum coke. Gasification is increasingly used with this feedstock (often using gasifiers placed in the refineries themselves).

Calcined

Calcined petroleum coke (CPC) is the product from calcining petroleum coke. This coke is the product of the coker unit in a crude oil refinery. The calcined petroleum coke is used to make anodes for the aluminium, steel and titanium smelting industry. The green coke must have sufficiently low metal content to be used as anode material. Green coke with this low metal content is called anode-grade coke. When green coke has excessive metal content, it is not calcined and is used as fuel-grade coke in furnaces.

Desulfurization

A high sulfur content in petcoke reduces its market value, and may preclude its use as fuel due to restrictions on sulfur oxides emissions for environmental reasons. Methods have thus been proposed to reduce or eliminate the sulfur content of petcoke. Most of them involve the desorption of the inorganic sulfur present in the pores or surface of the coke, and the partition and removal of the organic sulfur attached to the aromatic carbon skeleton.

Potential petroleum desulfurization techniques can be classified as follows:

- Solvent extraction.

- Chemical treatment.

- Thermal desulfurization.

- Desulfurization in an oxidizing atmosphere.

- Desulfurization in an atmosphere of sulfur-bearing gas.

- Desulfurization in an atmosphere of hydrocarbon gases.

- Hydrodesulfurization.

As of 2011 there was no commercial process available to desulfurize petcoke.

Storage, Disposal and Sale

Nearly pure carbon, petcoke is a potent source of carbon dioxide if burned.

Petroleum coke may be stored in a pile near an oil refinery pending sale. For example, in 2013 a large stockpile owned by Koch Carbon near the Detroit River was produced by a Marathon Petroleum refinery in Detroit which had begun refining bitumen from the oil sands of Alberta in November

2012. Large stockpiles of petcoke also existed in Canada as of 2013, and China and Mexico were markets for petcoke exported from California to be used as fuel. As of 2013 Oxbow Corporation, owned by William I. Koch, was a major dealer in petcoke, selling 11 million tons annually.

In 2017 a quarter of US exports of the fuel went to India, an Associated Press investigation found. In 2016 this amounted to more than eight million metric tons, more than 20 times as much as in 2010. India's Environmental Pollution Control Authority tested imported petcoke in use near New Delhi, and found sulfur levels 17 times the legal limit.

The International Convention for Prevention of Pollution from Ships (MARPOL 73/78), adopted by the IMO, has mandated that marine vessels shall not consume residual fuel oils (bunker fuel, etc) with a sulfur content greater than 0.1% from the year 2020. Nearly 38% of residual fuel oils are consumed in the shipping sector. In the process of converting excess residual oils into lighter oils by coking processes, pet coke is generated as a byproduct. Pet coke availability is expected to increase in the future due to falling demand for residual oil. Pet coke is also used in methanation plants to produce synthetic natural gas, etc. in order to avoid a pet coke disposal problem.

Health Hazards

Petroleum coke is sometimes a source of fine dust, which can penetrate the filtering process of the human airway, lodge in the lungs and cause serious health problems. Studies have shown that petroleum coke itself has a low level of toxicity and there is no evidence of carcinogenicity.

Petroleum coke can contain vanadium, a toxic metal. Vanadium was found in the dust collected in occupied dwellings near the petroleum coke stored next to the Detroit River. Vanadium is toxic in tiny quantities, 0.8 micrograms per cubic meter of air, according to the EPA.

Environmental Hazards

Environmental concerns stem from the storage and combustion of petcoke. By-waste accumulates as petcoke is processed, making waste management an issue. Petcoke's high silt content of 21.2% increases the risk of fugitive dust drifting away from petcoke mounds under heavy wind. An estimated 100 tons of petcoke fugitive dust including PM10 and PM2.5 are released into the atmosphere per year in the United States. Waste management and release of fugitive dust is especially an issue in the cities of Chicago, Detroit and Green bay.

Externalities stem from petcoke that cause potential environmental impacts. Petcoke is composed of 90% elemental carbon by weight which is converted to CO_2 during combustion. Use of petcoke also produces emissions of sulfur, and the potential for water pollution through nickel and vanadium runoff from refining and storage.

Lubricating Oil

Lubricating oil, sometimes simply called lubricant/lube, is a class of oils used to reduce the friction, heat, and wear between mechanical components that are in contact with each other. Lubricating oil is used in motorized vehicles, where it is known specifically as motor oil and transmission fluid.

There are two basic categories of lubricating oil: mineral and synthetic. Mineral oils are lubricating oils refined from naturally occurring crude oil. Synthetic oils are lubricating oils that are manufactured. Mineral lubricating oils are currently the most commonly used type because of the low cost of extracting the oils from crude oil. Additionally, mineral oils can be manufactured to have a varying viscosity, therefore making them useful in a wide range of applications.

Lubricating oils of different viscosities can be blended together, and it is this ability to blend them that makes some oils so useful. For example, common motor oil - shown in figure - is generally a blend of low viscosity oil to allow for easy starting at cool temperatures and a high viscosity oil for better performance at normal running temperatures.

Lubricating motor oil.

Types of Lubricants

In 1999, an estimated 37,300,000 tons of lubricants were consumed worldwide. Automotive applications dominate, but other industrial, marine, and metal working applications are also big consumers of lubricants. Although air and other gas-based lubricants are known (e.g., in fluid bearings), liquid lubricants dominate the market, followed by solid lubricants.

Lubricants are generally composed of a majority of base oil plus a variety of additives to impart desirable characteristics. Although generally lubricants are based on one type of base oil, mixtures of the base oils also are used to meet performance requirements.

Mineral Oil

The term "mineral oil" is used to refer to lubricating base oils derived from crude oil. The American Petroleum Institute (API) designates several types of lubricant base oil:

- Group I – Saturates < 90% and/or sulfur > 0.03%, and Society of Automotive Engineers (SAE) viscosity index (VI) of 80 to 120. Manufactured by solvent extraction, solvent or catalytic dewaxing, and hydro-finishing processes. Common Group I base oil are 150SN (solvent neutral), 500SN, and 150BS (brightstock).

- Group II – Saturates > 90% and sulfur < 0.03%, and SAE viscosity index of 80 to 120. Manufactured by hydrocracking and solvent or catalytic dewaxing processes. Group II base oil has superior anti-oxidation properties since virtually all hydrocarbon molecules are sat-urated. It has water-white color.

- Group III – Saturates > 90%, sulfur < 0.03%, and SAE viscosity index over 120 Manufactured by special processes such as isohydromerization. Can be manufactured from base oil or slax wax from dewaxing process.

- Group IV – Polyalphaolefins (PAO).

- Group V – All others not included above, such as naphthenics, polyalkylene glycols (PAG), and polyesters.

The lubricant industry commonly extends this group terminology to include:

- Group I+ with a viscosity index of 103–108.

- Group II+ with a viscosity index of 113–119.

- Group III+ with a viscosity index of at least 140.

Can also be classified into three categories depending on the prevailing compositions:

- Paraffinic.

- Naphthenic.

- Aromatic.

Synthetic Oils

Petroleum-derived lubricant can also be produced using synthetic hydrocarbons (derived ultimately from petroleum), "synthetic oils".

These include:

- Polyalpha-olefin (PAO).

- Synthetic esters.

- Polyalkylene glycols (PAG).

- Phosphate esters.

- Alkylated naphthalenes (AN).

- Silicate esters.

- Ionic fluids.

- Multiply alkylated cyclopentanes (MAC).

Solid Lubricants

PTFE: polytetrafluoroethylene (PTFE) is typically used as a coating layer on, for example, cooking utensils to provide a non-stick surface. Its usable temperature range up to 350 °C and chemical inertness make it a useful additive in special greases. Under extreme pressures, PTFE powder or solids is of little value as it is soft and flows away from the area of contact. Ceramic or metal or alloy lubricants must be used then.

Inorganic solids: Graphite, hexagonal boron nitride, molybdenum disulfide and tungsten disulfide are examples of solid lubricants. Some retain their lubricity to very high temperatures. The use of some such materials is sometimes restricted by their poor resistance to oxidation (e.g., molybdenum disulfide degrades above 350 °C in air, but 1100 °C in reducing environments.

Metal/alloy: Metal alloys, composites and pure metals can be used as grease additives or the sole constituents of sliding surfaces and bearings. Cadmium and gold are used for plating surfaces which gives them good corrosion resistance and sliding properties, Lead, tin, zinc alloys and various bronze alloys are used as sliding bearings, or their powder can be used to lubricate sliding surfaces alone.

Aqueous Lubrication

Aqueous lubrication is of interest in a number of technological applications. Strongly hydrated brush polymers such as PEG can serve as lubricants at liquid solid interfaces. By continuous rapid exchange of bound water with other free water molecules, these polymer films keep the surfaces separated while maintaining a high fluidity at the brush–brush interface at high compressions, thus leading to a very low coefficient of friction.

Biolubricant

Biolubricants are derived from vegetable oils and other renewable sources. They usually are triglyceride esters (fats obtained from plants and animals. For lubricant base oil use, the vegetable derived materials are preferred. Common ones include high oleic canola oil, castor oil, palm oil, sunflower seed oil and rapeseed oil from vegetable, and tall oil from tree sources. Many vegetable oils are often hydrolyzed to yield the acids which are subsequently combined selectively to form specialist synthetic esters. Other naturally derived lubricants include lanolin (wool grease, a natural water repellent).

Whale oil was a historically important lubricant, with some uses up to the latter part of the 20th century as a friction modifier additive for automatic transmission fluid.

In 2008, the biolubricant market was around 1% of UK lubricant sales in a total lubricant market of 840,000 tonnes/year.

Functions of Lubricants

One of the single largest applications for lubricants, in the form of motor oil, is protecting the internal combustion engines in motor vehicles and powered equipment.

Lubricant vs. Anti-tack Coating

Anti-tack or anti-stick coatings are designed to reduce the adhesive condition (stickiness) of a given material. The rubber, hose, and wire and cable industries are the largest consumers of anti-tack products but virtually every industry uses some form of anti-sticking agent. Anti-sticking agents differ from lubricants in that they are designed to reduce the inherently adhesive qualities of a given compound while lubricants are designed to reduce friction between any two surfaces.

Keep Moving Parts Apart

Lubricants are typically used to separate moving parts in a system. This separation has the benefit of reducing friction, wear and surface fatigue, together with reduced heat generation, operating noise and vibrations. Lubricants achieve this in several ways. The most common is by forming a physical barrier i.e., a thin layer of lubricant separates the moving parts. This is analogous to hydroplaning, the loss of friction observed when a car tire is separated from the road surface by moving through standing water. This is termed hydrodynamic lubrication. In cases of high surface pressures or temperatures, the fluid film is much thinner and some of the forces are transmitted between the surfaces through the lubricant.

Reduce Friction

Typically the lubricant-to-surface friction is much less than surface-to-surface friction in a system without any lubrication. Thus use of a lubricant reduces the overall system friction. Reduced friction has the benefit of reducing heat generation and reduced formation of wear particles as well as improved efficiency. Lubricants may contain additives known as friction modifiers that chemically bind to metal surfaces to reduce surface friction even when there is insufficient bulk lubricant present for hydrodynamic lubrication, e.g. protecting the valve train in a car engine at startup.

Transfer Heat

Both gas and liquid lubricants can transfer heat. However, liquid lubricants are much more effective on account of their high specific heat capacity. Typically the liquid lubricant is constantly circulated to and from a cooler part of the system, although lubricants may be used to warm as well as to cool when a regulated temperature is required. This circulating flow also determines the amount of heat that is carried away in any given unit of time. High flow systems can carry away a lot of heat and have the additional benefit of reducing the thermal stress on the lubricant. Thus lower cost liquid lubricants may be used. The primary drawback is that high flows typically require larger sumps and bigger cooling units. A secondary drawback is that a high flow system that relies on the flow rate to protect the lubricant from thermal stress is susceptible to catastrophic failure during sudden system shut downs. An automotive oil-cooled turbocharger is a typical example. Turbochargers get red hot during operation and the oil that is cooling them only survives as its residence time in the system is very short (i.e. high flow rate). If the system is shut down suddenly (pulling into a service area after a high-speed drive and stopping the engine) the oil that is in the turbo charger immediately oxidizes and will clog the oil ways with deposits. Over time these deposits can completely block the oil ways, reducing the cooling with the result that the turbo charger experiences total failure, typically with seized bearings. Non-flowing lubricants such as greases and pastes are not effective at heat transfer although they do contribute by reducing the generation of heat in the first place.

Carry away Contaminants and Debris

Lubricant circulation systems have the benefit of carrying away internally generated debris and external contaminants that get introduced into the system to a filter where they can be removed.

Lubricants for machines that regularly generate debris or contaminants such as automotive engines typically contain detergent and dispersant additives to assist in debris and contaminant transport to the filter and removal. Over time the filter will get clogged and require cleaning or replacement, hence the recommendation to change a car's oil filter at the same time as changing the oil. In closed systems such as gear boxes the filter may be supplemented by a magnet to attract any iron fines that get created.

It is apparent that in a circulatory system the oil will only be as clean as the filter can make it, thus it is unfortunate that there are no industry standards by which consumers can readily assess the filtering ability of various automotive filters. Poor automotive filters significantly reduces the life of the machine (engine) as well as making the system inefficient.

Transmit Power

Lubricants known as hydraulic fluid are used as the working fluid in hydrostatic power transmission. Hydraulic fluids comprise a large portion of all lubricants produced in the world. The automatic transmission's torque converter is another important application for power transmission with lubricants.

Protect against Wear

Lubricants prevent wear by keeping the moving parts apart. Lubricants may also contain anti-wear or extreme pressure additives to boost their performance against wear and fatigue.

Prevent Corrosion

Many lubricants are formulated with additives that form chemical bonds with surfaces or that exclude moisture, to prevent corrosion and rust. It reduces corrosion between two metallic surface and avoids contact between these surfaces to avoid immersed corrosion.

Seal for Gases

Lubricants will occupy the clearance between moving parts through the capillary force, thus sealing the clearance. This effect can be used to seal pistons and shafts.

Fluid Types

Automotive

- Engine oils:
  - Petrol (Gasolines) engine oils.
  - Diesel engine oils.
- Automatic transmission fluid.
- Gearbox fluids.
- Brake fluids.
- Hydraulic fluids.

Tractor (one Lubricant for All Systems)

- Universal Tractor Transmission Oil – UTTO.

- Super Tractor Oil Universal – STOU – includes engine.

Other Motors

- 2-stroke engine oils.

Industrial

- Hydraulic oils.

- Air compressor oils.

- Food Grade lubricants.

- Gas Compressor oils.

- Gear oils.

- Bearing and circulating system oils.

- Refrigerator compressor oils.

- Steam and gas turbine oils.

Aviation

- Gas turbine engine oils.

- Piston engine oils.

Marine

- Crosshead cylinder oils.

- Crosshead Crankcase oils.

- Trunk piston engine oils.

- Stern tube lubricants.

"Glaze" Formation (High Temperature Wear)

A further phenomenon that has undergone investigation in relation to high temperature wear prevention and lubrication, is that of a compacted oxide layer glaze formation. Such glazes are generated by sintering a compacted oxide layer. Such glazes are crystalline, in contrast to the amorphous glazes seen in pottery. The required high temperatures arise from metallic surfaces sliding against each other (or a metallic surface against a ceramic surface). Due to the elimination of metallic contact and adhesion by the generation of oxide, friction and wear is reduced. Effectively, such a surface is self-lubricating.

As the "glaze" is already an oxide, it can survive to very high temperatures in air or oxidising environments. However, it is disadvantaged by it being necessary for the base metal (or ceramic) having to undergo some wear first to generate sufficient oxide debris.

Disposal and Environmental Impact

It is estimated that 40% of all lubricants are released into the environment. Common disposal methods include recycling, burning, landfill and discharge into water, though typically disposal in landfill and discharge into water are strictly regulated in most countries, as even small amount of lubricant can contaminate a large amount of water. Most regulations permit a threshold level of lubricant that may be present in waste streams and companies spend hundreds of millions of dollars annually in treating their waste waters to get to acceptable levels.

Burning the lubricant as fuel, typically to generate electricity, is also governed by regulations mainly on account of the relatively high level of additives present. Burning generates both airborne pollutants and ash rich in toxic materials, mainly heavy metal compounds. Thus lubricant burning takes place in specialized facilities that have incorporated special scrubbers to remove airborne pollutants and have access to landfill sites with permits to handle the toxic ash.

Unfortunately, most lubricant that ends up directly in the environment is due to general public discharging it onto the ground, into drains and directly into landfills as trash. Other direct contamination sources include runoff from roadways, accidental spillages, natural or man-made disasters and pipeline leakages.

Improvement in filtration technologies and processes has now made recycling a viable option (with rising price of base stock and crude oil). Typically various filtration systems remove particulates, additives and oxidation products and recover the base oil. The oil may get refined during the process. This base oil is then treated much the same as virgin base oil however there is considerable reluctance to use recycled oils as they are generally considered inferior. Base-stock fractionally vacuum distilled from used lubricants has superior properties to all natural oils, but cost effectiveness depends on many factors. Used lubricant may also be used as refinery feedstock to become part of crude oil. Again, there is considerable reluctance to this use as the additives, soot and wear metals will seriously poison/deactivate the critical catalysts in the process. Cost prohibits carrying out both filtration (soot, additives removal) and re-refining (distilling, isomerisation, hydrocrack, etc.) however the primary hindrance to recycling still remains the collection of fluids as refineries need continuous supply in amounts measured in cisterns, rail tanks.

Occasionally, unused lubricant requires disposal. The best course of action in such situations is to return it to the manufacturer where it can be processed as a part of fresh batches.

Environment: Lubricants both fresh and used can cause considerable damage to the environment mainly due to their high potential of serious water pollution. Further the additives typically contained in lubricant can be toxic to flora and fauna. In used fluids the oxidation products can be toxic as well. Lubricant persistence in the environment largely depends upon the base fluid, however if very toxic additives are used they may negatively affect the persistence. Lanolin lubricants are non-toxic making them the environmental alternative which is safe for both users and the environment.

Liquefied Petroleum Gas

Liquefied petroleum gas or LPG is a type of hydrocarbon gas that is obtained by refining crude oil or processing natural gas. This gas is composed of either propane and butane by themselves or as a mixture of the two. In addition to its use as a fuel for cooking and heating, LPG is also important for use in manufacturing applications, as a fuel for cars, and it can be used to power cogeneration plants.

An LPG cylinder.

Production

Liquefied petroleum gas is produced during the refining process of crude oil or extracted during the processing of natural gas. The gases produced in this process are mainly propane and butane with small amounts of other gases. These gases are liquefied through pressurization to make them easier to transport and store.

To liquefy the fuel, gases are stored in sturdy tanks and held at high pressures—about 20 times atmospheric pressure. These tanks have additional safety features because of this extreme pressurization, mainly a built-in shutoff valve to seal the tank if there are leaks and an extra-sturdy design. Since LPG generally has no odour, small amounts of ethanethiol (a foul smelling mercaptan, a type of odorant) are added to help people smell dangerous gas leaks.

Use

LPG has a high caloric value, meaning that it is a good energy source as it provides a high level of heat. It is also a valuable fuel as it has almost no sulfur content, which results in cleaner burning. About half of the LPG used is consumed for heating and cooking and essentially is used in place of natural gas. The remaining 50% of LPG is split more or less equally between use in cars and industrial uses. Overall, LPG provides less than 2% of the total energy people use but it is still a major alternative to gasoline.

When used, LPG is generally delivered by trucks in a large tank and placed outside a home or other building. In addition, reusable gas canisters are available for powering stoves, heaters, and barbecues. Small canisters of LPG are also available for portable hair styling tools.

Two major disadvantages of the use of LPG are safety and cost. The high pressure needed to store LPG results in occasional tank bursts if canisters are not stored properly and maintained. In addition, LPG is highly flammable. However, suppliers take many safety precautions to ensure that LPG is as safe as ordinary natural gas supply. The cost of LPG is several times higher than ordinary natural gas, but LPG might prove to be a good option if access to natural gas is not available.

Use as a Vehicle Fuel

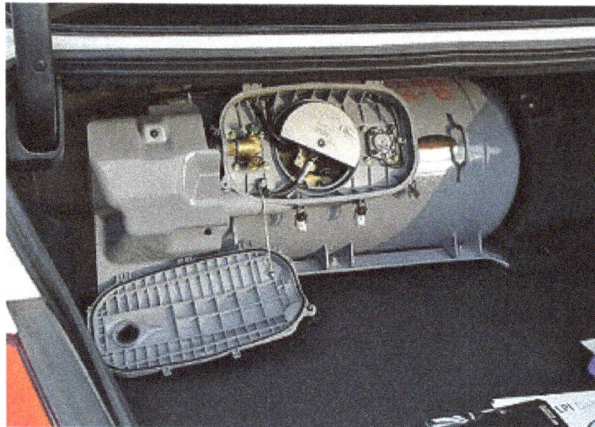

A vehicle's LPG tank.

Liquefied petroleum gas can be used as an alternative fuel to power internal combustion engines as it is more cleanly burning than gasoline and can produce lower amounts of some harmful emissions such as carbon dioxide. An estimated 6 million European vehicles run on LPG and there are around 17,500 gasoline stations in Europe supplying this fuel.

One of the major benefits of using LPG as an alternative fuel is that it is usually less expensive than gasoline. Additionally, the careful design and safety features of the tanks make them slightly safer to use than gasoline simply because they have a shut-off valve. This minimizes the risk of an LPG fire if used as a vehicle fuel.] LPG can also be used as a fuel without taking away from vehicle performance.] The use of LPG also helps with the issue of importing fuels for use from other countries as 90% of LPG used in the US comes from domestic sources.

However, the availability of vehicles that are LPG-fueled is limited. Some existing vehicles can be converted to use this LPG with installations, but involves installing a separate fuel system as the liquid is stored in highly pressurized fuel tanks. In addition, it is harder to find places to refill LPG as it is not as widely used as gasoline or diesel, and fewer miles can be traveled on a single tank of LPG.

Propane

Propane is a three-carbon alkane, normally a gas, but compressible to a liquid that is transportable. It is derived from other petroleum products during oil or natural gas processing. It is commonly used as a heat source for engines, barbecues, and homes. Its name was derived from propionic acid When commonly sold as fuel, it is also known as liquified petroleum gas (LPG or LP-gas) and can be a mixture of propane with smaller amounts of propylene, butane, and butylene. The odorant (ethanethiol) is also added so that people can easily smell the gas in case of a leak. In North America, LPG is primarily propane (at least 90 percent), with the rest mostly butane and propylene. This is the HD5 standard, primarily written for vehicle fuels; note that not all product labeled "propane" conforms to this standard.

Sources of Propane

Propane is not produced for its own sake, but as a byproduct of two other processes: natural gas processing and petroleum refining.

The processing of natural gas involves removal of propane and butane from the natural gas, to prevent condensation of these liquids in natural gas pipelines. Additionally, oil refineries produce some propane as a by-product of production of gasoline or heating oil.

The supply of propane cannot be easily adjusted to account for increased demand because of the by-product nature of propane production. About 85 percent of U.S. propane is domestically produced.

The United States imports about 10-15 percent of the propane consumed each year. Propane is imported into the United States via pipeline and rail from Canada, and by tankers from Algeria, Saudi Arabia, Venezuela, Norway, and the United Kingdom.

After it is produced, North American propane is stored in huge salt caverns located in Fort Saskatchewan, Alberta, Canada, Mont Belvieu, Texas, and Conway, Kansas. These salt caverns were hollowed out in the 1940s and can store up to 80 million barrels of propane, if not more. When the propane is needed, most of it is shipped by pipelines to other areas of the Midwest, the North, and the South, for use by customers. Propane is also shipped by barge and rail car to selected U.S. areas.

Properties and Reactions

Propane undergoes combustion reactions in a similar fashion to other alkanes. In the presence of excess oxygen, propane burns to form water and carbon dioxide.

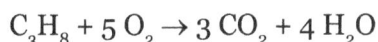

$$C_3H_8 + 5\ O_2 \rightarrow 3\ CO_2 + 4\ H_2O$$

When not enough oxygen is present for complete combustion, propane burns to form water and carbon monoxide.

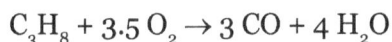

$$C_3H_8 + 3.5\ O_2 \rightarrow 3\ CO + 4\ H_2O$$

Unlike natural gas, propane is heavier than air (1.5 times denser). In its raw state, propane sinks and pools at the floor. Liquid propane will flash to a vapor at atmospheric pressure and appears white due to moisture condensing from the air.

When properly burned, propane produces about 2,500 BTU of heat per cubic foot of gas.

Propane is nontoxic; however, when abused as an inhalant it poses a mild asphyxiation risk through oxygen deprivation. It must also be noted that commercial product contains hydrocarbons beyond propane, which may increase risk. Propane and its mixtures may cause frostbite during rapid expansion.

Propane combustion is much cleaner than gasoline, though not as clean as natural gas. The presence of C-C bonds, plus the C=C bond of propylene, create organic exhausts besides carbon dioxide and water vapor. These bonds also cause propane to burn with a visible flame.

Uses

Retail sale of propane cylinders.

It is used as fuel in cooking on many barbecues and portable stoves, and in motor vehicles. The ubiquitous, 5-gallon steel container has been dubbed the "barbecue bottle." Propane powers some locomotives, buses, forklifts, and taxis and is used for heat and cooking in recreational vehicles and campers. In many rural areas of North America, propane is also used in furnaces, stoves, water heaters, laundry dryers, and other heat-producing appliances. 6.5 million American households use propane as their primary heating fuel. Also recently, Tippmann, a paintball company, has made a paintball gun called the "C3." This gun's propellent is propane as opposed to the usual carbon dioxide or nitrogen.

Domestic and Industrial Fuel

In North America, local delivery trucks called "bobtails" fill up large tanks that are permanently installed on the property (sometimes called pigs), or other service trucks exchange empty bottles of propane with filled bottles. The bobtail is not unique to the North American market, though the practice is not as common elsewhere, and the vehicles are generally referred to as tankers. In many countries, propane is delivered to consumers via small or medium-sized individual tanks.

Retail sale of propane in Monmouth, Oregon.

Propane is the fastest growing fuel source in the Third World, especially in China and India. Its use frees up the huge rural populations from time-consuming ancient chores such as wood gathering and allows them more time to pursue other activities, such as increased farming or educational opportunties. Hence, it is sometimes referred to as "cooking gas."

On an aside, North American barbecue grills powered by propane cannot be used overseas. The "propane" sold overseas is actually a mixture of propane and butane. The warmer the country, the higher the butane content, commonly 50/50 and sometimes reaching 75 percent butane. Usage is calibrated to the different-sized nozzles found in non-U.S. grills. Americans who take their grills overseas — such as military personnel — can find U.S.-specification propane at AAFES military post exchanges.

North American industries using propane include glass makers, brick kilns, poultry farms, and other industries that need portable heat. Additionally, most of the entire North American chemical industry uses propane to power their huge facilities that crack or distill industrial chemical products.

Refrigeration

Propane is also instrumental in providing off-the-grid refrigeration, also called gas absorption refrigerators. Made popular by the Servel company, propane-powered refrigerators are highly efficient, do not require electricity, and have no moving parts. Refrigerators built in the 1930s are still in regular use, with little or no maintenance. However, certain Servel refrigerators are subject to a recall for CO poisoning.

In highly purified form, propane (R-290) can serve as a direct replacement in mechanical refrigeration systems designed to use R-12, R-22, or R-134a chloro- or fluorocarbon based refrigerants. Today, the Unilever Ice Cream company and others are exploring the use of environmentally friendly propane as a refrigerant. As an added benefit, users are finding that refrigerators converted to use propane are 9-15 percent more energy efficient.

Vehicle Fuel

Propane is also being used increasingly for vehicle fuels In the U.S., 190,000 on-road vehicles use propane, and 450,000 forklifts use it for power. It is the third most popular vehicle fuel in America, behind gasoline and diesel. In other parts of the world, propane used in vehicles is known as autogas. About nine million vehicles worldwide use autogas.

The advantage of propane is its liquid state at room temperature. This allows fast refill times, affordable fuel tank construction, and ranges comparable to (though still less than) gasoline. Meanwhile it is noticeably cleaner, results in less engine wear (due to carbon deposits) without diluting engine oil (often extending oil-change intervals), and until recently was a relative bargain in North America. However, public filling stations are still rare. Many converted vehicles have provisions for topping off from "barbecue bottles." Purpose-built vehicles are often in commercially owned fleets, and have private fueling facilities.

Propane is generally stored and transported in steel cylinders as a liquid with a vapor space above the liquid. The vapor pressure in the cylinder is a function of temperature. When gaseous propane is drawn at a high rate the latent heat of vaporization required to create the gas will cause the bottle to cool (this is why water often condenses on the sides of the bottle and then freezes). In extreme cases this may cause such a large reduction in pressure that the process can no longer be supported. In addition, the lightweight, high-octane compounds vaporize before the heavier, low-octane ones. Thus, the ignition properties change as the tank empties. For this reason, the liquid is often withdrawn using a dip tube.

Other

- Propane is also used as a feedstock for the production of base petrochemicals in steam cracking.

- It is used in some flamethrowers, as the fuel, or as the pressurizing gas.

- Some propane becomes a feedstock for propyl alcohol, a common solvent.

- It is used as fuel in hot air balloons.

- It is used as a propellant along with silicon (for lubrication) in airsoft guns.

Butane

A butane torch for kitchen use
(specifically for crème brûlée).

Butane is an an alkane with the chemical formula C_4H_{10}. As a type of hydrocarbon, it can undergo hydrocarbon combustion which gives off heat energy. Butane is one of the hydrocarbon components of raw natural gas, which is a type of fossil fuel. Butane is usually removed from natural gas before being shipped to customers, but then butane is sold separately as a fuel itself.

Butane is commonly mixed with propane in camping fuel in order to maintain higher pressures at low temperatures. Butane is also one of the main components in lighter fluid and is commonly used in cigarette lighters, portable stoves and butane torches. Figure shows a butane torch used for cooking purposes.

Properties

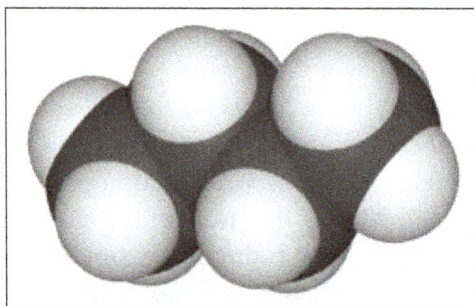

Space filling model of butane, the white spheres
representhydrogen and the black spheres represent carbon.

Table of some of the basic properties of butane.

Formula	C_4H_{10}
Molar mass	58.12 grams/mole
Energy density	49.5 MJ/kg
Melting Point	-138°C
Boiling Point	-0.5°C

Combustion Reaction

Butane releases its chemical energy by undergoing hydrocarbon combustion. Below is a hydrocarbon combustion animation showing the net reaction that occurs when butane combines with oxygen.

$$2\ (C_4H_{10}) + 13\ (O_2) \rightarrow 8(CO_2) + 10(H_2O) + \text{Heat Energy (Enthalpy)}$$

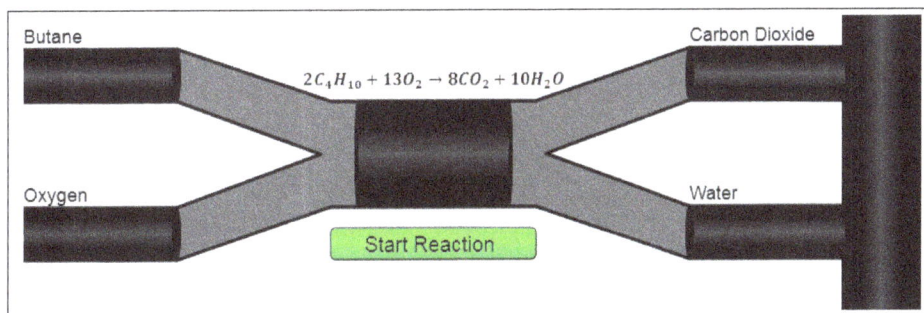

The hydrocarbon combustion reaction releases heat energy and is an example of an exothermic reaction. The reaction also has a negative enthalpy change (ΔH) value.

Uses of Butane Gas

Butane Torch

This is an item that takes advantage of its flammable properties. The butane torch most commonly uses it for glass making, craft projects, and certain plumbing projects which require heat. Butane can be used in portable grills meaning campers can make good use of it. In this case, the butane is stored in a gas canister. Due to its ability to easily compress, gas canisters are a great choice for storing butane.

LPG

Butane can be combined with propane as well as other substances in order to form liquefied petroleum gas, which is also known as LPG. Liquefied petroleum gas is found in the manufacturing of petrochemicals that are used in its purest state, and it can also be used for calibrating instruments and as a refrigerant.

Refrigerators

Very pure forms of butane can be used as refrigerants and due to the risk methane places on the ozone layer, it has replaced the use of methane in household refrigerators. Additionally, adding butane to gasoline doesn't increase the flammability of the gasoline, but enhances its performance and quality. This hydrocarbon is used as a food additive too.

Lighters and Aerosols

Butane is often used as the fuel in lighters as it can handle being pressurised. As the vapour pressure for butane is relatively low, putting it in a small plastic pressure vessel such as a lighter is possible. One of their most common distinguishing characteristics is that they can be compressed into a liquid at relatively low pressure.

When the pressure is released through the valve, it turns instantly to gas which is very easily ignited. Being cheap and a hydrocarbon, it does not attack plastics – these characteristics result in a compound which perfectly fulfils the properties needed for a small, safe lighter

Petroleum Jelly

Petroleum jelly is also called Petrolatum translucent, yellowish to amber or white, unctuous substance having almost no odour or taste, derived from petroleum and used principally in medicine and pharmacy as a protective dressing and as a substitute for fats in ointments and cosmetics. It is also used in many types of polishes and in lubricating greases, rust preventives, and modeling clay.

Petrolatum is obtained by dewaxing heavy lubricating-oil stocks. It has a melting-point range from 38° to 54 °C (100° to 130 °F). Chemically, petrolatum is a mixture of hydrocarbons, chiefly of the paraffin series.

Naphtha

Naphtha is a name given to several mixtures of liquid hydrocarbons that are extremely volatile and flammable. Each such mixture is obtained during the distillation of petroleum or coal tar, and occasionally by the distillation of wood. Accordingly, it is known by different names, such as petroleum naphtha, coal-tar naphtha, or wood naphtha.

Naphtha is used primarily as feedstock for producing a high-octane gasoline component via the catalytic reforming process. It is also used in the petrochemical industry for producing olefins in steam crackers and in the chemical industry for solvent (cleaning) applications.

Properties

To obtain the product known as *naphtha,* a complex soup of chemicals is broken into another range of chemicals, which are then graded and isolated mainly by their specific gravity and volatility. As a result, the product contains a range of distinct chemicals with a range of properties. They generally have a molecular weight range of 100-215, a specific gravity range of 0.75-0.85, and a boiling point range of 70-430 °F. Their vapor pressure is usually less than 5 mm mercury.

Naphthas are insoluble in water. They are colorless (with a kerosene odor) or red-brown (with an aromatic odor). They are incompatible with strong oxidizers.

Generally speaking, less dense naphthas ("light naphthas") have a higher paraffin content. They are therefore also called paraffinic naphtha. The denser naphthas ("heavy naphthas") are usually richer in naphthenes and aromatics, and they are therefore referred to as N&A's.

Production of Naphtha in Refineries and Uses

Naphtha is obtained in petroleum refineries as one of the intermediate products from the distillation of crude oil. It is a liquid intermediate between the light gases in the crude oil and the heavier liquid kerosene. Naphthas are volatile, flammable and have a specific gravity of about 0.7. The generic name naphtha describes a range of different refinery intermediate products used in different applications. To further complicate the matter, similar naphtha types are often referred to by different names.

The different naphthas are distinguished by:

- Density (g/ml or specific gravity).

- PONA, PIONA or PIANO analysis, which measures (usually in volume percent but can also be in weight percent):

    ◦ Paraffin content (volume percent).

    ◦ Isoparaffin content (only in a PIONA analysis).

    ◦ Olefins content (volume percent).

    ◦ Naphthenes content (volume percent).

    ◦ Aromatics content (volume percent).

Paraffinic (or Light) Naphthas

The main application for paraffinic ("light") naphthas is as feedstock in the petrochemical production of olefins. This is also the reason they are sometimes referred to as "light distillate feedstock" or LDF. (These naphtha types may also be called "straight run gasoline" (SRG) or "light virgin naphtha" (LVN).)

When used as feedstock in petrochemical steam crackers, the naphtha is heated in the presence of water vapor and the absence of oxygen or air until the hydrocarbon molecules fall apart. The primary products of the cracking process are olefins (ethylene / ethene, propylene / propene and butadiene) and aromatics (benzene and toluene). These are used as feedstocks for derivative units that produce plastics (polyethylene and polypropylene, for example), synthetic fiber precursors (acrylonitrile), and industrial chemicals (glycols, for instance).

Heavy Naphthas

The "heavy" naphthas can also be used in the petrochemical industry, but they are more often used as feedstock for refinery catalytic reformers where they convert the lower octane naphtha to a higher octane product called reformate. Alternative names for these types are "straight run benzene" (SRB) or "heavy virgin naphtha" (HVN).

Additional Applications

Naphthas are also used in other applications, such as:

- In the production of gasoline.
- In industrial solvents and cleaning fluids.
- An oil painting medium.
- The sole ingredient in the home cleaning fluid energine, which has been discontinued. You can purchase this type of naphtha at any hardware store.
- An ingredient in shoe polish.
- An ingredient in some lighter fluids for wick type lighters such as zippo lighters.
- An adulterant to petrol.
- A fuel for portable stoves and lanterns, sold in north america as white gas or coleman fuel.
- Historically, as a probable ingredient in greek fire (together with grease, oil, sulfur, and naturally occurring saltpeter from the desert).
- A fuel for fire spinning, fire juggling, or other fire performance equipment which creates a brighter and cleaner yet shorter burn.
- To lightly wear the finish off guitars when preparing "relic" instruments.
- To remove oil from the aperture blades of camera lenses, which if present can slow the movement of the blades, leading to overexposure.

Health and Safety Considerations

Forms of naphtha may be carcinogenic, and products sold as naphtha frequently contain some impurities that may have deleterious properties of their own. Given that the term *naphtha* is applied to different products, each containing a variety of distinct chemicals, it is difficult to make rigorous comparisons and to identify specific carcinogens. This task is further complicated by the presence of a number of other known and potential carcinogens in modern environments.

Below are links to some Material Safety Data Sheet (MSDS) specifications for different "naphtha" products, which contain varying proportions of naphtha and other chemicals. Besides giving health guidelines, they provide one of the few ways to determine what a given product contains:

- Petroleum Ether MSDS.

- Diggers Australia Shellite.

- Shell Ronsonol Lighter Fuel.

- MSDS for camping-stove fuels.

Benzene in particular is a known high-risk carcinogen, and so benzene content is typically specified in the MSDS. But more specific breakdown of particular forms of hydrocarbon is not as common.

According to J. LaDou in Occupational and Environmental Medicine, "Almost all volatile, lipid-soluble organic chemicals cause general, nonspecific depression of the central nervous system or general anesthesia." The U.S. Occupational Health and Safety Administration (OSHA) places the permissible exposure limit (PEL) at 100 parts per million (ppm); and the Health Hazards/Target Organs are listed as eyes, skin, RS, CNS, liver, and kidney. Symptoms of acute exposure are dizziness and narcosis with loss of consciousness. The World Health Organization categorizes health effects into three groups: reversible symptoms (Type 1), mild chronic encephalopathy (Type 2) and severe chronic toxic encephalopathy (Type 3).

Toxicity

Toxicity dose response exposures may be impacted (decreased or increased) by chemical, biological, and environmental factors:

- Chemical factors include concentrations of the chemicals, their interactions with one another, dispersability, toxicity, water solubility, particle size, bioavailability, persistence in the body, and so forth.

- Biological factors include stress, respiratory rate, gender, age, race, individual susceptibility, route of entry, rate of uptake, storage in the body, metabolism, and excretion.

- Environmental factors can affect chemical and particulate exposures, such as by temperature, air pressure, air quality, and precipitation.

Air Sampling for Naphthas

Air sampling is conducted to identify and evaluate employee or source exposures of potentially hazardous gases or particulates; assess compliance; and evaluate process or reformulation changes.

Two categories of air sampling equipment exist, they are: direct reading and sample collection:

- Direct reading equipment provides immediate measurement of exposure concentration.

- Sample collection equipment takes samples of air over a period of time, and these samples are then weighed and analyzed in a laboratory.

Sample collection involves active and passive air monitoring methods. Active sampling relies on sampling pumps to draw air and chemical vapors or gases to adsorbent filter materials. Passive monitors rely on the collection of gases and vapors through passive diffusion to allow personal sampling without use of pumps.

Sampling Types

Various types of sampling may be used, as noted below:

- Personal sampling: Personal sampling is used to evaluate employee exposure to naphtha. The employee wears the sampling device that collects an air sample representative of air exposure for a specific period of time.

- Area Sampling: Area Sampling is used to evaluate background exposure to leaks and implement control measures.

- Grab Sampling: Grab sampling is used to monitor extremely toxic environments over a short period of time, or to determine if additional air monitoring is required for over-exposure.

- Integrated Sampling: Integrated exposure sampling is used to determine the 8-hour time weighted average exposure because various exposure concentrations are integrated during the sampling period.

Complications

Complications with air sampling can occur in the form of interference with chemicals (alcohols, ketones, ethers, and halogenated hydrocarbons), vapors, sampling media, humidity, temperature, barometric pressure, atmospheric dust, water vapor, and container.

Exposure Control

Primary methods focus on preventing chemical exposures before they occur. Personal protective equipment could include the use of air-purifying cartridges, respirators, and gloves. Engineering prevention controls would include automated handling, enclosure and elimination of harmful substances, isolation, and change of process. Ventilation controls would include local exhaust ventilation and vacuum operations. Administrative prevention controls would include changes in work practices, education, training, job rotation, job reduction, job reassignment, and proper maintenance and housekeeping.

Secondary methods focus on early identification and treatment of chemical exposures.

Tertiary methods include the treatment and rehabilitation of employees overexposed to harmful chemicals in the workplace.

References

- Petroleum-products-and-by-products, conferences-list: omicsonline.org, Retrieved 14 July 2019

- Bodén, Roger. "Paraffin Microactuator" (PDF). Materials Science Sensors and Actuators. University of Uppsala. Archived from the original (PDF) on 8 February 2012. Retrieved 25 October 2013

- Naphtha, entry: newworldencyclopedia.org, Retrieved 08 April 2019

- Seager, Spencer L.; Slabaugh, Michael (19 January 2010). "Alkane reactions". Chemistry for Today: General, Organic, and Biochemistry. Belmont, California: Cengage. p. 364. ISBN 978-0-538-73332-8

- Petroleum-jelly, science: britannica.com, Retrieved 24 August 2019

- "U.S. world's biggest supplier of heavy oil refining byproduct". Star-Advertiser. Honolulu. Associated Press. December 1, 2017. Retrieved December 1, 2017

- Popular-uses-butane-gas: adamsgas.co.uk, Retrieved 14 July 2019

Permissions

Index

www.ingramcontent.com/pod-product-compliance
Lightning Source LLC
Chambersburg PA
CBHW082051190326
41458CB00010B/3508